Powder and High-Solid Coatings

Powder and High-Solid Coatings

Editor

Stefano Rossi

MDPI • Basel • Beijing • Wuhan • Barcelona • Belgrade • Manchester • Tokyo • Cluj • Tianjin

Editor
Stefano Rossi
University of Trento
Italy

Editorial Office
MDPI
St. Alban-Anlage 66
4052 Basel, Switzerland

This is a reprint of articles from the Special Issue published online in the open access journal *Coatings* (ISSN 2079-6412) (available at: https://www.mdpi.com/journal/coatings/special_issues/powder_solid_coat).

For citation purposes, cite each article independently as indicated on the article page online and as indicated below:

LastName, A.A.; LastName, B.B.; LastName, C.C. Article Title. *Journal Name* **Year**, *Volume Number*, Page Range.

ISBN 978-3-0365-4719-0 (Hbk)
ISBN 978-3-0365-4720-6 (PDF)

Contents

About the Editor

Stefano Rossi

Stefano Rossi is head of the Coatings and Industrial Anticorrosion laboratory, Department of Industrial Engineering of the University of Trento. He holds a degree in Material Engineering from the University of Trento (1991) and a PhD in Materials Science and Technology from the University of Florence (1997). He is a teacher of the courses of "Protection of Materials and Structures" and "Product Design" at the University of Trento and "Materials Science and Technologies" at the Free University of Bozen. His research topics are the protection of materials from corrosion, in particular with the use of coatings (organic, metallic and ceramic), and their monitoring using electrochemical techniques and accelerated laboratory tests. In addition to the corrosion and abrasion behavior of the surfaces, he is interested in the aesthetic-perceptive properties (color, gloss). The research is also aimed at the study of smart and innovative coatings and layers. He has more than 180 articles in international journals.

coatings

MDPI

Editorial

Powder and High-Solid Coatings

Stefano Rossi

Department of Industrial Engineering, University of Trento, Via Sommarive 9, 38123 Trento, Italy; stefano.rossi@unitn.it; Tel.: +39-0461-282442

Abstract: This Special Issue presents a series of research papers and reviews about the actual trend of powder and high solid coatings which show the advantage of great environmental sustainability by avoiding the massive use of organic solvents. Moreover, some very interesting studies exist on the move from a simple protective layer to a smart coating with additional properties, both for the aesthetic and functional aspects.

Keywords: corrosion protection; graphene; aluminum trihydrate; pearlescent pigments; marine atmosphere; rust converter; colloidal silver; scanning electrochemical microscopy; electrochemical tests; laboratory accelerated test

Organic coatings, often referred to as paint or varnish, are the most popular method of protecting metal substrates from corrosion. Thanks to their excellent protective properties, organic coatings have been applied increasingly more often over the years.

Let us introduce some basic information on organic coatings.

This type of coating consists of four elements. First, a binder, which constitutes the polymer film and is responsible for adhesion with the metal substrate and the cohesion of pigments and additives. We then have pigments or insoluble solid particles with different purposes, such as corrosion inhibition, providing color, modifying the surface roughness and creating surface effects. The third element, additives, are chemical compounds or particles of organic or inorganic type which are added to modify the properties of paints, such as anti-foamers, stabilizers and rheology modifiers, and, fillers to reduce the cost of the organic coating. Finally, the solvent, present in many painted systems but not in all and used to keep the paint in a liquid state during storage and application. The binder and pigments are solubilized or dispersed in the solvent.

We will later see the gradual evolution of the painted systems which are the theme and purpose of this Special Issue.

Paints can be classified according to the type of binder. Each type of binder has characteristics that make it more or less suitable for certain uses and environments; for example, applying it as a top coat or a primer.

The epoxy matrix is excellent as a primer due to its corrosion resistance and adhesion properties. Otherwise, it is less suitable for use as a top coat due to its poor UV resistance. On the other hand, acrylic binders have excellent resistance to aging and abrasion. At the same time, they possess a surface appearance that makes them ideal candidates as finishing layers. Polyester resins are also interesting for this last use as they have good mechanical properties together with a low cost. Conversely, the cost penalizes the polyurethane-type binders which, due to their valuable properties of resistance to aging and abrasion, combined with adhesion and aesthetic properties would be very attractive.

There are different types of paints. The oldest and still most widely used are based on organic solvents and therefore present environmental issues. The organic coating (binder, pigments and additives) that dissolves in the solvent is what forms the protective layer following the evaporation of the solvent, hence the title of this Special Issue.

In fact, to reduce the environmental impact, risk of fire and potential threat to the operator's safety, alternatives to organic solvents have been sought more and more frequently.

Citation: Rossi, S. Powder and High-Solid Coatings. *Coatings* **2022**, *12*, 786. https://doi.org/10.3390/coatings12060786

Received: 19 May 2022
Accepted: 31 May 2022
Published: 7 June 2022

Water-based paints initially showed a reduced use, mainly due to their limited protection performance compared to traditional solvent-based paints, but they have experienced a tremendous level of development in recent years.

With the aim of reducing the use of solvents as much as possible or eliminating them altogether, the use of high solid paints is widespread in North America. On the other hand, in Europe, powder paints are preferred to solve environmental problems. In this case, a coating is applied to the substrate starting from solid particles in the form of a powder, which contain both the binder, the pigments and the additives. By applying an electric field, suitably bonded particles are attracted to the substrate. By adhering to the substrate, being insulators, they favor the deposition of subsequent particles in the areas of substrate that remain uncovered. By means of a curing heat treatment, the chains cross-link and the protective coating is obtained.

Regarding the use of powder paints, scientists have often wondered if some components, additives or pigments that are used in liquid paints can also be used in this type of coating. The transfer is not immediate since a substance that is soluble in an organic solvent is unlikely to be easily dispersed in a solid matrix. Moreover, there is a risk of not obtaining uniform dispersion but rather the formation of accumulations and clusters. It is therefore necessary to carry out a bonding treatment so that the different components comprising the paint (matrix, pigments and additives) are simultaneously present in a single grain of powder. Even in the instance of managing to homogeneously disperse the components, there is still the question of whether the additives and pigments show the same behavior in powder paint as they do traditional solvent-based paint [1,2]. Hence, the article by Weihong Li et al. "Investigation of the Performance of ATH Powders in Organic Powder Coatings" [3], which evaluates the optimization and behavior of aluminum trihydrate (ATH), one of the most widely used fillers in solvent-borne coatings and polymer industries, introduced in a powder system.

One of the positive aspects of paints is their possibility of being deposited on different types of metal substrates, such as steel, cast iron, aluminum and copper alloys. Paints find endless potential applications on products with different functions requiring use in the most varied environments, not just in non-aggressive indoor environments or in natural environments. In this case, to ensure the expected service life it is necessary to carefully evaluate the aggressiveness of the environment and the required protection time to consider the possibility of mechanical damage to the coating and the possibility of carrying out maintenance operations.

One of the most critical applications for a protective system is the protection of marine and offshore structures. The marine atmosphere is a challenge for the designer, due to the high humidity, the presence of aggressive ions such as chlorides, the high insolation and the difficulty of restoration [4,5]. Usually, for these applications, spray-deposited solvent-based paint systems consisting of several layers and a very high thickness are used. Turning to powder and high solid coatings for environmental reasons, the question arises as to whether these types of coatings can guarantee performances that are similar to those of traditional solvent-based paints. Thus, the review by Krystel Pélissier and Dominique Thierry entitled, "Powder and High-Solid Coatings as Anticorrosive Solutions for Marine and Offshore Applications? A Review" [6] is very interesting, as it illustrates all the important aspects that are necessary to obtain corrosion protection with these types of paints in high-performance applications. Exposure tests in very aggressive natural environments (Florida test or Arizona test) or accelerated laboratory tests, such as continuous or cycled salt spray, are traditionally used to evaluate the performance of a painted system; however, these tests present some critical issues. They often take a very long time, from months up to several years, to establish whether the painted system shows the desired protective performance. Electrochemical techniques, and in particular Electrochemical Impedance Spectroscopy (EIS) [7], help scientists by allowing a reduction in evaluation times and providing many data that enable a better understanding of the protection mechanism of a coating, as well as the evaluation of the reduction in protective properties over time. In addition to traditional electrochemical techniques that require a fairly large test area, in recent years, localized electrochemical techniques have been developed and used that enable greater depth of study of the corrosion and protection mechanisms by analyzing very small samples at a microscopic level. Here fits the work of Ricardo M. Souto and co-authors, entitled, "Uses of Scanning Electrochemical Microscopy (SECM) for the Characterization with Spatial and Chemical resolution of Thin Surface Layers and Coating Systems Applied on Metals: A Review" [8]. As mentioned, scientific research in the field of paints is aimed at reducing the environmental impact by seeking alternative solutions that are less polluting and more efficient. For this reason, innovative pigments, new additives and green corrosion inhibitors are constantly being studied. In recent years, graphene G and graphene oxide GO, consisting of thin sheets of sp2 hybridized carbon, have also aroused great interest in the field of paints [9–11]. However, the dispersion of graphene-based flakes is not simple and is susceptible to clustering.

An example of the introduction of graphene oxide in a painted system is present in the work by Carmen Perez and coauthors, entitled, "Influence of Graphene Oxide Additions on the Corrosion Resistance of a Rust Converter Primer" [12]. To study the effects of flakes on the protective properties of primers, the GO was included in five rust converters, identifying the optimal percentage between the converter and the GO sheets.

In recent years, organic coatings have also found further applications that are different from the traditional corrosion protection function of metal substrates. Recently, coatings that offer other characteristics or that respond to an external stimulus with a different or unexpected response have been developed and studied. These materials represent the so-called smart coatings. The introduction of graphene was also aimed at obtaining an organic coating with a high surface conductivity. Other examples may be the introduction of NIR near infrared reflective pigments to produce coatings for roof applications that allow the temperature of buildings to be passively controlled, with the aim of reducing the effects of urban heat islands.

Applying organic coatings to surfaces that possess antimicrobial features in order to reduce the necessity of a regular disinfection has also gained great interest following the COVID-19 pandemic. One of the most effective methods is through the introduction of silver particles into the paints, thanks to the recognized antimicrobial action of this element [13,14]. Some questions remain; for example, how much paint is needed to make the coating effective without an excessive increase in the cost of production or a reduction in the other properties of the paint? There is also the need to disperse the silver as uniformly as possible and ensure its presence in the outermost layers of the deposit so that the antimicrobial action can be implemented. This topic is discussed in the article entitled, "Durability of Acrylic Cataphoretic Coatings Additivated with Colloidal Silver" [15].

Finally, let us revisit one of the main reasons that lead paints to be the most used type of coating: color. Normally, in scientific research, no attention is paid to color. No reviewer asks the authors about the color of the paints they are studying if this has not been indicated. Very little research is aimed at evaluating the color of organic coatings. Studies are usually limited to the engineering properties of the coatings (corrosion resistance, abrasion resistance, hardness, electrical and thermal insulation, thermal characteristics); however, color is a key aspect. The first thing we notice about a product is its shape and surface, its texture and color. We desire a specific color for our car and a great many other objects., The world would be a sad, black and white (or rather gray, green and brown) place if we only thought about the corrosion protection properties of a paint. We all have the red Ferrari, the gray Porsche, Hermès orange, and other iconic colors in mind.

In recent years, always with the aim of customizing and diversifying the product, new pigments have been introduced that produce optical effects. Metallic, micaceous and pearlescent are the possibilities offered by some organic pigments [16,17]. The pearlescent effect is obtained with pigments that have a superficial oxide layer with different thicknesses from area to area so that an interference with light enables the observer to perceive different colors. Once only used in niche sectors, now more and more of these pigments are employed in common products, hence the need to evaluate not only the corrosion protection properties but also the perceptive aesthetic features of the paints. Today, it is important to be able to maintain the surface appearance of organic coatings over time even in very aggressive environments that are subjected to solar radiation. This topic is covered in the latest paper of the Special Issue entitled, "Study of the Durability and Aesthetical Properties of Powder Coatings Admixed with Pearlescent Pigments" [18].

To conclude, we can say that paints that are made from simple organic layers with toxic pigments (such as red lead) and sad colors have evolved and continue to evolve, presenting properties that were once unexpected, both from a technical and aesthetic point of view.

We can confidently say that the times when "Any customer can have a car painted any color that he wants, as long as it is black" are definitely a memory.

Funding: This publication received no external funding.

Acknowledgments: As Editor of this Special Issue, I would like to thank first of all the authors of the articles who have shown an interest in these research topics, but also the reviewers, editors and all those who have contributed to the publication of this Special Issue.

Conflicts of Interest: This author declares no conflict of interest.

References

1. Fernández-Álvarez, M.; Velasco, F.; Bautista, A.; Galiana, B. Functionalizing organic powder coatings with nanoparticles through ball milling for wear applications. *Appl. Surf. Sci.* **2020**, *513*, 145834. [CrossRef]
2. Rossi, S.; Fedel, M.; Deflorian, F.; Zanol, S. Influence of different colour pigments on the properties of powder deposited organic coatings. *Mater. Des.* **2013**, *50*, 332–341. [CrossRef]
3. Li, W.; Franco, D.C.; Yang, M.S.; Zhu, X.; Zhang, H.; Shao, Y.; Zhang, H.; Zhu, J. Investigation of the Performance of ATH Powders in Organic Powder Coatings. *Coatings* **2019**, *9*, 110. [CrossRef]
4. Olajire, A.A. Recent advances on organic coating system technologies for corrosion protection of offshore metallic structures. *J. Mol. Liq.* **2018**, *269*, 572–606. [CrossRef]
5. Yu, M.; Fan, C.; Ge, F.; Lu, Q.; Wang, X.; Cui, Z. Anticorrosion behavior of organic offshore coating systems in UV, salt spray and low temperature alternation simulated Arctic offshore environment. *Mater. Today Commun.* **2021**, *28*, 102545. [CrossRef]
6. Pélissier, K.; Thierry, D. Powder and High-Solid Coatings as Anticorrosive Solutions for Marine and Offshore Applications? A Review. *Coatings* **2020**, *10*, 916. [CrossRef]
7. Margarit-Mattos, I.C.P. EIS and organic coatings performance: Revisiting some key points. *Electrochim. Acta* **2020**, *354*, 136725. [CrossRef]
8. Santana, J.J.; Izquierdo, J.; Souto, R.M. Uses of Scanning Electrochemical Microscopy (SECM) for the Characterization with Spatial and Chemical Resolution of Thin Surface Layers and Coating Systems Applied on Metals: A Review. *Coatings* **2022**, *12*, 637. [CrossRef]
9. Calovi, M.; Rossi, S.; Deflorian, F.; Dirè, S.; Ceccato, R. Effect of functionalized graphene oxide concentration on the corrosion resistance properties provided by cataphoretic acrylic coatings. *Mater. Chem. Phys.* **2020**, *239*, 121984. [CrossRef]
10. Calovi, M.; Rossi, S.; Deflorian, F.; Dirè, S.; Ceccato, R.; Guo, X.; Frankel, G.S. Effects of Graphene-Based Fillers on Cathodic Delamination and Abrasion Resistance of Cataphoretic Organic Coatings. *Coatings* **2020**, *10*, 602. [CrossRef]
11. Zhang, X.; Ma, R.; Du, A.; Liu, Q.; Fan, Y.; Zhao, X.; Wu, J.; Cao, X. Corrosion resistance of organic coating based on polyhedral oligomeric silsesquioxane-functionalized graphene oxide. *Appl. Surf. Sci.* **2019**, *484*, 814–824. [CrossRef]
12. Díaz, B.; Nóvoa, X.R.; Pérez, C.; Rodríguez-Morgado, M. Influence of Graphene Oxide Additions on the Corrosion Resistance of a Rust Converter Primer. *Coatings* **2022**, *12*, 345. [CrossRef]
13. Sharaf, M.H.; Nagiub, A.M.; Salem, S.S.; Kalaba, M.H.; El Fakharany, E.M.; Abd El-Wahab, H. A new strategy to integrate silver nanowires with waterborne coating to improve their antimicrobial and antiviral properties. *Pigment Resin Technol.* **2022**, *51*. [CrossRef]
14. Palza, H.; Delgado, K.; Curotto, N. Synthesis of copper nanostructures on silica-based particles for antimicrobial organic coatings. *Appl. Surf. Sci.* **2015**, *357*, 86–90. [CrossRef]
15. Calovi, M.; Rossi, S. Durability of Acrylic Cataphoretic Coatings Additivated with Colloidal Silver. *Coatings* **2022**, *12*, 486. [CrossRef]
16. Rohrer, A.; Venturini, M.T. Pearlescent Pigments in Coatings—A Primer. *Paint Coat. Ind.* **2018**, *34*, 10, 30–34.

17. Kneiphof, T.; Klein, R. Real-time image-based lighting of metallic and pearlescent car paints. *Comput. Graph.* **2022**, *105*, 36–45. [CrossRef]
18. Rossi, S.; Russo, F.; Bouchakour Rahmani, L. Study of the Durability and Aesthetical Properties of Powder Coatings Admixed with Pearlescent Pigments. *Coatings* **2020**, *10*, 229. [CrossRef]

coatings

MDPI

Article

Investigation of the Performance of ATH Powders in Organic Powder Coatings

Weihong Li [1], Diego Cárdenas Franco [2], Marshall Shuai Yang [2], Xinping Zhu [3], Haiping Zhang [1], Yuanyuan Shao [1], Hui Zhang [1,2,*] and Jingxu Zhu [1,2]

[1] Collaborative Innovation Center of Chemical Science and Engineering (Tianjin), School of Chemical Engineering and Technology, Tianjin University, Tianjin 300072, China; weihli@tju.edu.cn (W.L.); hpzhang@tju.edu.cn (H.Z.); yshao@tju.edu.cn (Y.S.)

[2] Particle Technology Research Center, Department of Chemical & Biochemical Engineering, The University of Western Ontario, London, ON N6A 3K7, Canada; cardenas_franco@yahoo.ca (D.C.F.); marshall.yang@uwo.ca (M.S.Y.); jzhu@uwo.ca (J.Z.)

[3] Wesdon-Rivers Institute of Powder Coatings Science and Technology, Zhaoqing 526000, China; xzhu269@gmail.com

* Correspondence: hzhang1@tju.edu.cn; Tel.: +86-22-23497607

Received: 7 January 2019; Accepted: 5 February 2019; Published: 12 February 2019

Abstract: Aluminum trihydrate (ATH) is one of the most widely used fillers in solvent-borne coatings and polymer industries. To investigate its performance in powder coatings, four different pigment to binder ratios (P/B) were used in respect to different ATH concentrations in four most widely used resin systems (epoxy, epoxy-polyester, polyurethane and polyester). The angle of repose (AOR) and specific gravity of powder paints increase accordingly with ATH concentration. The mechanical properties, corrosion and UV resistance performances of coating films are enhanced or only slightly decreased with the addition of ATH. Pencil scratch hardness increase significantly with the increase of ATH content. Impact resistance decreases within 5% of the initial resistance at the maximum ATH loading. There is a slightly decrease in corrosion resistance performances of less than 2 mm of failure at scribe in all formulations. The incorporation of ATH has slight effect on the UV resistance performance of all the samples. Moreover, the matting effect of ATH is observed for all the resin systems, especially epoxy, with the specular gloss decreasing more than 70% at P/B of 0.75.

Keywords: ATH; organic powder coatings; polymer matrix

1. Introduction

Due to the economic viability and the ease of application and storage of them, powder coatings are considered as an advanced and modern coating system [1]. In contrast to solvent-borne coatings, powder coatings have zero-volatile organic content (VOC). It is estimated that in the current coating systems, the recyclability of coating powder which does not adhere to the substrates is up to 98% [2,3]. Thus, they could offer a highly promising approach to produce eco-friendly coatings. Meanwhile, when taking into account the low cost of waste disposal and rejection rate, less costly maintenance and fast curing, the overall operating cost of a powder application plant is much lower than that of traditional solvent-borne coatings [4,5].

Powder coatings have five principal components that include polymer resin, curing agent, pigment, additive and filler. In general, the polymer resin and curing agent play the most critical roles because the combination determines the mechanical performance and durability in the final application of powder coatings. Pigment, additive and filler are also very important components of powder coatings. Pigments have the aesthetic purpose of giving color to the paint. Additives are designed to modify the performance of the coating at low concentrations, usually up to 2% of the formulation by weight. Fillers are normally inorganic minerals that are chemically inert and produced artificially or naturally. The addition of fillers serves two purposes: the first is to reduce the content of the polymer resin which is the most expensive part of the formulation since they are derived from petroleum; the second is to modify certain mechanical properties of the coating [6–10].

The most common fillers used in powder coating formulations include blanc fixe ($BaSO_4$), lithopone ($ZnS{\cdot}BaSO_4$), talc ($Mg_3Si_4O_{10}(OH)_2$), zinc white (ZnO), dolomite ($CaMg(CO_3)_2$) and carbon black (C) etc. Generally, including fillers in the paint, the hardness increases, causing a slight decrease in its flexibility, as well as toughness and ultimate elongation [11–14]. Besides their commonness, each filler also has its unique effect. For example, the advantages of lithopone include its excellent hiding power and brightness, and reduced cost [15]. Talc also has a high hiding power, a matting effect and give a satin finish [16]. Zinc white is a popular filler used as a UV stabilizer and has a relatively high refractive index which makes it an efficient white pigment [17]. Investigating the overall performance of fillers in coating systems provides valuable information on the selection of fillers.

Aluminum trihydrate (ATH) is a type of popular filler in liquid coating. It is the purified byproduct of the first stage of the production process of aluminum metal and can be produced in a variety of particle sizes that are controlled by grinding or precipitation processes. Due to its high thermal conductivity, it can be blended with RTV (room temperature vulcanized) silicone rubber coatings to improve the tracking and erosion resistance [18]. Furthermore, ATH works as the most widely used flame retarder. It begins to release water of hydration that absorbs heat required for the gasification and Al_2O_3 is generated correspondingly. This action, the cooling effect, is an important way to remove heat from the hot spots [19,20]. On the other hand, the formation of the thermally stable ceramic Al_2O_3 can prevent the escaping of some of the volatile species from entering the gas phase and consequently prevents the evolution of toxic or corrosive substances [21]. Besides its wide application in solvent-borne coatings and polymer industries, several ATH manufactures and suppliers, such as Huber materials Inc., have also paid attention to the superior performance of ATH on improving the properties of powder coatings. Yet, little systematic research is so far regarding its effects on the physical and mechanical properties, UV and corrosion resistance performances in powder coatings and especially ultrafine powder coatings. Ultrafine powder coatings ($D_{50} < 25$ μm) can produce a smoother and thinner film which leads to improved visual appearance and considerable cost savings [1]. Compared with regular powder coatings ($D_{50} > 30$ μm), the ultrafine powder coatings are more sensitive to the side-effects brought over by fillers due to their poorer powder flowabilities and thinner film build. Therefore, ultrafine powder coatings are used in this study to present and understand the performance of ATH in powder coatings.

In this study, ATH was incorporated in the four most widely used powder coating resin systems, epoxy (E), epoxy-polyester (H), polyurethane (PU) and polyester (PE). The effects of ATH concentration on flowability of ultrafine powder paints, physical and mechanical properties of finished films as well as their performance in corrosive environments and exterior conditions were carefully investigated.

2. Experimental

2.1. Preparation of the Samples

As described in the introduction, powder coatings have five principal components that include polymer resin, curing agent, pigment, additive and filler. The first two are the most basic parts of powder paint formulation, called "Binder". The other vital parts, the pigment, additive and filler, are called "Pigment" in general. Each one of the components has its own role in directly affecting the properties and performance of the paint. In this study, the materials used are shown in Table 1. The pigment to binder ratio (P/B ratio or P/B) is a common parameter to present the components of a powder coating. It can be expressed in terms of weight fraction as Equation (1).

$$P/B = \frac{\text{Weight of Pigments}}{\text{Weight of Binders}}, \quad (1)$$

Different samples were made by increasing the P/B. In general, the increased P/B will achieve a lower cost in resin systems while leading to a poorer crosslink inside coatings and between coatings and the substrates [22].

In this study, four different P/B ratios (0.028, 0.25, 0.5 and 0.75) were tested in each of the four resin systems. Correspondingly, the contents of ATH in formulations are 0 wt.%, 17.0 wt.%, 34.0 wt.% and 41.0 wt.% in respect to each P/B. To determine the effect of P/Bs, the type and content of pigments (1.4 wt.%), degassing agent (0.3 wt.%), as well as fluidization additives (0.7 wt.%) were kept constant. Physical properties of ATH powders used are shown in Table 2.

Table 1. Materials.

Materials	Epoxy (E)	Hybrid (H)	Polyurethane (PU)	Polyester (PE)
Resin	DER 663U	PE (Crylcoat 2440-2): E (DER 663U) = 7:3 wt.%	Albester 3220	Crylcoat 2440-2
Curing agent	DICY/Vestagon B31		IPDI & ε-caprolactam/Alcure 4400	TGIC/Araldite PT 810
Filler	Alumina trihydrate/Custom Grinders Polyfill 301			
Degassing agent	Benzoin/S602			
Pigment	Carbon black/Raven 5000 Ultra II			
Fluidization additive	Acrylic polymer/Lanco P10			

Table 2. Physical properties of aluminum trihydrate (ATH) powders.

Filler	Shape	Particle Size (D_{50})/[μm]	Specific Gravity/[-]	Color	Refractive Index/[-]	Oil Absorption/ [g oil/100 g]
$Al(OH)_3$	Irregular	9	2.4	White	1.58	28

In the manufacturing process of the sample powder coatings, the appropriate components are premixed and extruded with a twin-screw extruder (SLJ-10, Donghui Powder Coating Processing Equipment Co., Ltd., Yantai, China). The hot extrudates are allowed to cool down, crushed to chips, and then admitted into an Air classifier mill (ACM, ACM-02, Donghui Powder Coating Processing Equipment Co., Ltd.) for fine grinding, classifying and further screening. The particle size of the finished paint powders was 18 ± 0.5 μm (D_{50}), which was analyzed with a laser particle size analyzer (BT2000B, Bettersize instruments Ltd., Dandong, China).

Each one of the samples was applied on two types of panels: Aluminum panel (88.9 × 63.5 × 0.8 (L × W × T), mm^3, Q-Lab Corporation Ltd., Westlake, OH, USA) and Steel panel (127.0 × 76.2 × 0.8 (L × W × T), mm^3, Q-Lab Corporation Ltd.). All the panels were degreased with acetone, rinsed with distilled water and dried with nitrogen gas before spraying. The powder paints were sprayed with a corona gun (Sure coat, Nordson Corporation, Westlake, OH, USA). All the samples were subjected to a curing schedule of 10 min at 204 °C which guarantees that the coating is fully cured. After curing, at least two aluminum panels and two steel panels with a dry film thickness of 38.1 μm ± 5% were chosen to perform all the necessary tests. The thickness of the coating was measured with a thickness gage (PosiTector 6000, DeFelsko Corporation, Ogdensburg, NY, USA).

2.2. *Measurements of the Samples*

2.2.1. Characterization of Powder Paints

Measurement of angle of repose (AOR) is an efficient method to determine the flowability of a powder. The AOR is defined as the slope measured from the horizontal to the side of a cone that is formed when the powder is allowed to free fall onto a plate. It is the largest angle at which powders can pile up. According to the Cheremisinoff's theory [23], the AOR is related to the powders' cohesiveness and internal friction and is widely used to characterize flow properties of powders. As shown in Figure 1, the powders can be categorized into five groups, from very free-flowing to very cohesive [23].

In this sense, a powder sample with an AOR higher than 45° is normally considered cohesive while a sample with a lower AOR has better flowability. The AOR of powder paints was measured by a powder characteristics tester (PT-X, Hosokowa Micron Corporation, Hirakata, Japan).

The specific gravity is an important parameter that is taken to calculate the cost effectiveness of a powder coatings product, normally in the form of "square meter" coverage per kilogram of powder coatings at a given film thickness. Also, a powder coating with a higher specific gravity normally exhibits poorer flowability due to the stronger inter-particle forces [3,24]. The specific gravity of powder paints was measured following ASTM D5965-02(2013) [25].

In the experiment, all powder samples are produced by an air classifier mill with nearly the same operating parameters with minor adjustment to ensure all the powder samples are in the same particle shape and their D_{50}s are in the range of 18 ± 0.5 μm. Therefore, the difference of AORs caused by particle size and particle shape can be ignored.

2.2.2. Characterization of Finished Films

The performances of finished films depend upon many conditions, such as quality of resin systems and conditions exposed to. Many test methods for monitoring the performance of films are designed to simulate the conditions to which powder coatings are subjected. In order to evaluate the film properties such as hardness, impact resistance, flexibility, weathering resistance and corrosion resistance, suitable standards are used to specify the detailed procedures of evaluation.

Pencil scratch hardness test is the simplest coating hardness test. It uses special pencils with different degrees of hardness (from 9H to 9B) to scratch the coating and then determines its hardness. The test was carried out using a pencil scratch hardness tester (PH5800, BYK Additives and Instruments Ltd., Wesel, Germany), following ASTM D3363-05(2011) [26].

The Impact resistance of the coatings was evaluated using a variable height impact tester following D2794-93(2010) [27]. A steel weight with a hemispherical head (13mm in diameter) and an overall weight of 1.8kg was dropped from various heights (20–120 cm) onto the coated panels. If no cracks were evident, the same procedure was repeated increasing the height by 25 mm incremental until cracks appear, producing the failure of the coating. The maximum impact resistance value of the impact tester was 21.2 J.

Flexibility of the finished films was measured by conical mandrel bend test apparatus (BYK 5750, BYK Additives and Instruments Ltd.) following ASTM D522M-13 [28]. The coating panels were fixed vertically in the conical mandrel. The mandrel is then rolled up, which bends the coated panel into approximately 135°. The diameter of the cone varies from 38 mm to 3 mm. The coating is considered flexible if there is no crack on its surface.

Specular gloss is a measurement of the light reflectance of a test surface. It is determined by a gloss meter which directs a light at a specific angle to the test surface and measures the amount of reflection. The percentage of the light that is reflected at the same angle is measured by a photocell. The Specular gloss was measured using a gloss meter (IQ206085, Rhopoint Components Ltd., East Grinstead, UK) following ASTM D523-14(2018) [29].

The salt spray test is an important method in evaluating corrosion resistance performance of coating films under atmospheric conditions. The test was conducted using a salt spray chamber (MX-9204, Associated Environmental Systems Ltd., Hong Kong) following ASTM B117-2011 [30]. In the study, a scribe in X form was made with a cutting tool on the surface of the coated panel leaving the metal substrate exposed. Based on ASTM D1654-08(2016) [31], the representative mean rust creepage from scribe can be evaluated by a rating number varying from 10 to 1 grade, as shown in Table 3.

Table 3. Rating number of failures at scribe.

Rust Creepage/mm	Rating Number	Rust Creepage/mm	Rating Number
Zero	10	5.0 to 7.0	4
0 to 0.5	9	7.0 to 10.0	3
0.5 to 1.0	8	10.0 to 13.0	2
1.0 to 2.0	7	13.0 to 16.0	1
2.0 to 3.0	6	16.0 to more	0
3.0 to 5.0	5	–	–

The resistance of a powder coating to the UV radiation is one of the most significant parameters to determine how suitable a particular paint is for exterior conditions. The test chamber (XE-3, Q-Lab Corporation Ltd.) used in the study is equipped with xenon-arc lamps which have the maximum wavelength of 340 nm. A cycle consists of 102 min of UV irradiation (with radiation level of 0.35 W/m^2) at 63 °C in dry conditions and 18 min in water spray conditions (sprayed water for 20 s and stopped for 40 s). The panels were cut in halves, one for the test and the other as control. The measured specular gloss at 60° was recorded to calculate the gloss retention of the paint following ASTM D523-14(2018) [29]. The gloss retention is defined as Equation (2):

$$\text{Gloss retention} = \frac{\text{Gloss}_{t=x}}{\text{Gloss}_{t=0}} \quad (2)$$

The microstructure and morphology of finished films were analyzed with scanning electron microscopy (SEM, S4800, Hitachi High-Technologies Global, Tokyo, Japan). The chemical composition was studied via an energy-dispersive spectroscopy system (EDS, X-Max N, Oxford Instruments, Abingdon, UK).

All the above characterizations of the finished films were performed one day after cured.

3. Results

3.1. Powder Properties-Angle of Repose and Specific Gravity

Figure 1 shows the angle of repose values (AORs) of the paint powders blended with fluidization additives based on four resin systems with respect to P/B ratios. It can be seen that the type of base polymer resin has a significant influence on the AOR of a paint powder, with the hybrid having the lowest AORs at any P/B ratios followed by epoxy, polyurethane and polyester. On the other hand, for all the four resin systems, raising the ATH content consistently increases their AORs and thereby reduces their flowability. From the prior study and the authors' industrial experience [3], the paint powders with AORs higher than 42° tend to agglomerate and exhibit poor flow behavior on the spray lines during powder coating application. According to this standard, the flowabilities of most of the powders are acceptable after the addition of fluidization additives, except for those of the polyurethane and polyester at P/Bs equal to or over 0.5. For almost all formulations, the paint powders are fair to passable flow and expected to exhibit a good flow behavior during the electrostatic spraying process.

Figure 2 presents the specific gravities of the paint powders of four resin systems at different P/Bs. All the samples have similar specific gravities ranging from 1.20 to 1.23 when ATH are excluded. Evidently, the addition of ATH into the paint powders increases their specific gravities and the formulations having the same P/B have close values. As mentioned before, a higher powder coating specific gravity would normally correspond to a poorer flowability, the results in Figure 2 are in agreement with this expectation.

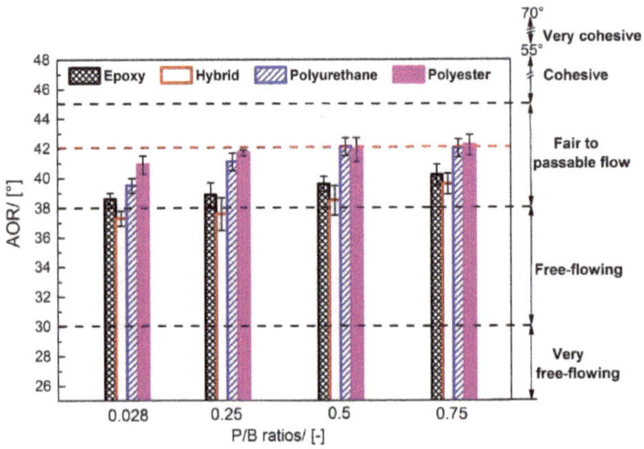

Figure 1. Angle of repose values (AORs) of paint powders with different resins and pigment to binder ratios (P/Bs).

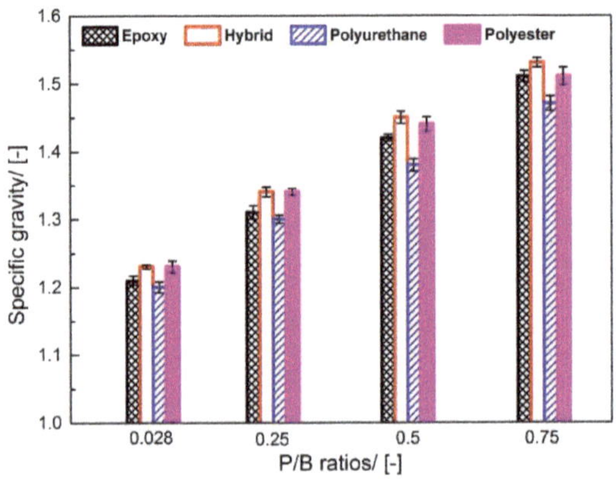

Figure 2. Specific gravities of formulations with different resins and P/Bs.

3.2. Film Properties-Mechanical, Physical, UV and Corrosion Resistance Performances

3.2.1. Effect of ATH on the Pencil Scratch Hardness, Flexibility and Impact Resistance of the Films

For typical protective coating films, high surface hardness, strong impact resistance and reliable flexibility to the base substrates are crucial requisites. Figure 3 shows the pencil scratch hardness test results of formulations based on the four resin systems with varied P/B ratio. As expected, the pencil scratch hardness is greatly influenced by initial hardness of the resin system and the inclusion of ATH. The epoxy-based formulations have the highest hardness of 2H without ATH (E-0.028), while other ATH-free formulations have the hardness of HB in contrast. Moreover, the addition of ATH has a positive effect on the pencil scratch hardness of finished films for all investigated resin systems. Increasing ATH loading (P/B) results in distinct improvement of pencil scratch hardness, which increase to 4H, 4H, 2H and H for E-0.75, PU-0.75, H-0.75 and PE-0.75, respectively. Besides, the growth of hardness is various between the panels based on different resin systems. For example, pencil scratch hardness of panels based on E, H and PE increase at P/Bs over 0.5, while those based on PU increase more or less in sync with the increase of P/B. Above all, the effect of ATH is most significant on PU with the hardness increasing from HB to 4H.

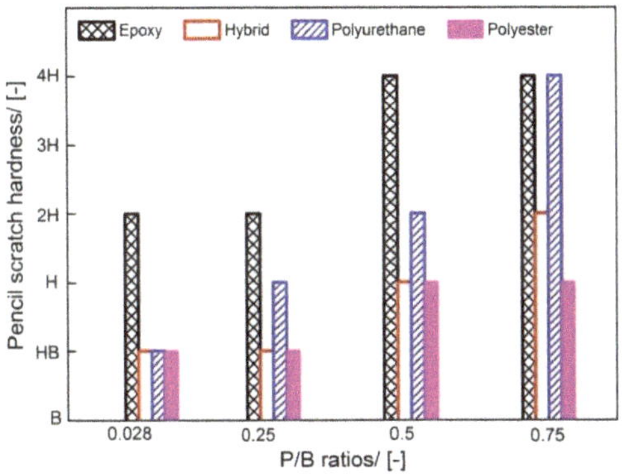

Figure 3. Pencil scratch hardness of formulations with different resins and P/Bs.

Flexibility of the powder coatings was evaluated in the conical mandrel bend test. At room temperature, samples demonstrated good flexibility up to P/B of 0.75, and when bent over a 3mm mandrel diameter (the lowest diameter of the cone used in the study), did not show any signs of cracking. Further addition of ATH in the powder coating may affect the flexibility of the paint film in a significant way, because the increasing amount of ATH creates discontinuities in the polymer matrix which reduces the overall flexibility of the coating. But the effect may be reduced by applying thinner coatings that can be achieved with ultrafine powder coatings which provide this clear advantage over thick regular size powder paints [1].

Figure 4 illustrates the comparison of the impact resistance for various coatings based on different resin systems and P/Bs. The impact resistance of the coatings was evaluated by an impact tester with different impact energies. As it was expected, the addition of inorganic fillers into the formulations caused a reduction in the impact resistance of the finished films. The discontinuities in the matrix make it less flexible and consequently lose adhesion to the substrate when it receives an impact. But the effect of ATH is not significant. Except for PU paints, the reduction of impact resistance only occurs at the highest P/B of 0.75, with the impact resistance reduces from 21.2 J to 20.1 J. While for PU paints, the effect of ATH starts at the P/B of 0.5.

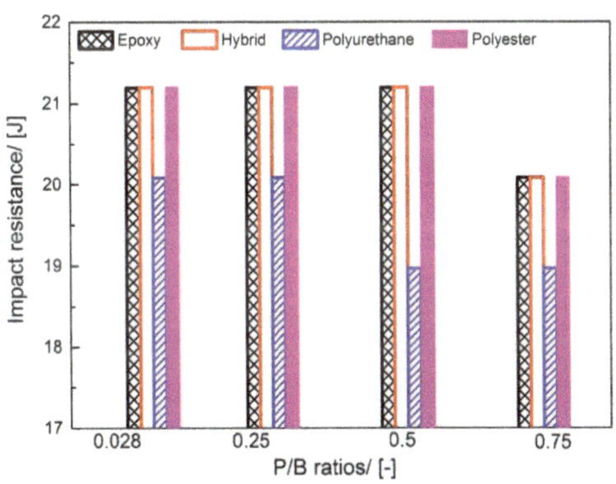

Figure 4. Impact resistance of formulations with different resins and P/Bs.

3.2.2. Effect of ATH on the Specular Gloss of the Films

Specular gloss is a function of light reflection, the more light is reflected and partially directed, the more glossy the surface appears to be. Figure 5 presents the results of specular gloss at 60° from the finished films at different P/Bs. When no ATH is added to the formulations, the specular gloss varies from 94–97, and quickly decreases with the addition of ATH. The increases of P/B reduce the specular gloss. At the addition of the maximum load of ATH (P/B of 0.75), specular gloss of hybrid coating decreases by 47.8% in contrast to ATH-free coatings, while it falls down more significantly for epoxy coatings, by 71.1%.

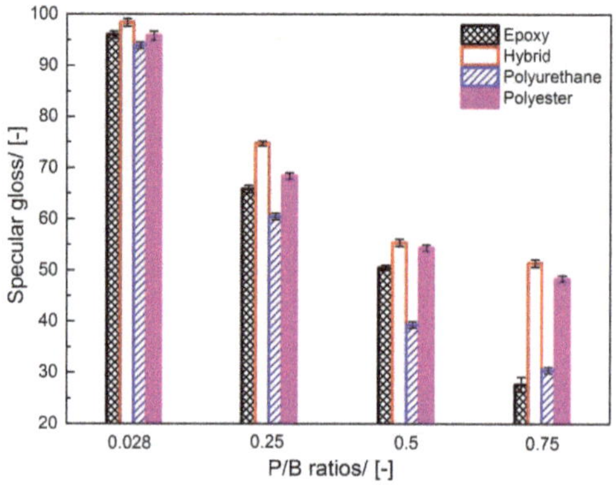

Figure 5. Specular gloss of formulations with different resins and P/Bs.

The decrease in specular gloss indicates an increase in matting effects. Generally, powder coatings are matted to a certain extent by the addition of matting agents. Matting agents are components that are not compatible with the resin, or have high oil absorption, and thus cause a micro texture on the surface which scatters the incident light in different directions making it less glossy. Usually, matting agents are polyolefinic waxes and inorganic extenders such as aerosol, silica and talc that can reduce the gloss level by about 60% of the original when used in concentrations between 2% and 4%. But the excess use of wax will produce haze or yellowing as well as an oily coating [11,12]. The use of ATH as fillers can directly enhance the matting effect without incorporating any other matting agents. The matting effect is achieved by decreasing specular gloss by about 50–70% of the initial gloss at the maximum ATH loading, which is comparable or even higher than that of common matting agents. The specular gloss reduction caused by addition of ATH is far more significant than the normal fillers such as $BaSO_4$, etc. In this sense, ATH is more suitable for semi to low-gloss coatings.

The excellent matting effect of ATH may be due to the floatation and high oil absorption of ATH particles in polymeric matrix which leads to the micro-roughness in the film surface. The formation of a micro-rough film surface structure is necessary for the diffuse light-scattering for creating the visual effect of reduced gloss [32,33]. It is important to mention that, ATH has a hydrophilic surface which results in poor interfacial compatibility and particle agglomeration in a polymeric matrix [34], and this would tend to make the ATH particles rise, during the melt-flow process (the first stage in the curing process), to the top-surface zone of the film and thus make some of them partially embedded in the films, as shown in Figure 6b. Besides, ATH powder has a relatively high oil absorption of 28 g oil/100 g in contrast to other popular fillers, such as blanc fixe ($BaSO_4$) and lithopone ($ZnS{\cdot}BaSO_4$), which have much lower oil absorption of 11–17 g oil/100 g and 11–14 g oil/100 g. Generally, the higher the oil absorption value of the filler, the higher the filler's surface area and the more binder it will require to bind it. The combination of high surface free energy and oil absorption leads to the increasing of the concentration of ATH particles in the top-surface zone of the film and therefore the formation of a micro-rough film surface.

The surface structure is identified by SEM images and EDS analysis. Sample H-0.028 shows a good dispersion and without many agglomerates in the surface (Figure 6a), whereas marked agglomerates could be found in the coatings H-0.75 incorporated with the maximum ATH loading, as shown in Figure 6b. The micro-rough surfaces produced by agglomerates result in a poorer corrosion resistance performance, which will be discussed in Section 3.2.4. Otherwise, elemental mapping of the coating films shows that there are high contents of carbon (C) and oxygen (O) in the surface of the coating, which originated from hybrid resin polymers and ATH powders (Figure 7c,d). ATH particles in Figure 7a are identified by the mapping of element aluminum (Al). It is seen that most of the aluminum appears in the surface of the coating film, as shown in Figure 7b.

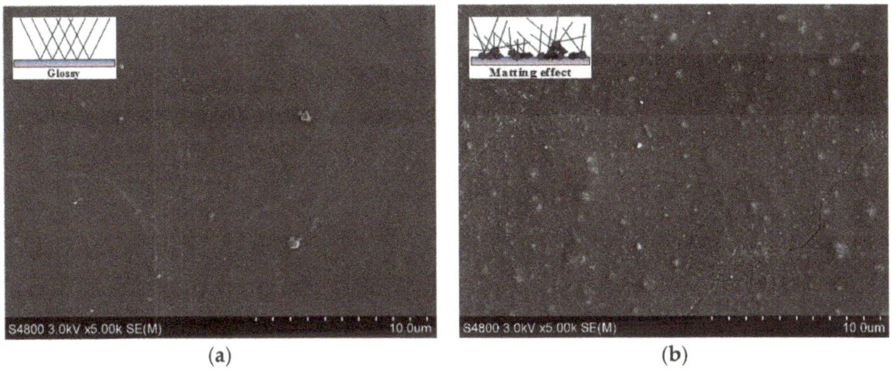

Figure 6. SEM micrographs of coating films with the addition of ATH. (**a**) Formulation H-0.028; (**b**) Formulation H-0.75.

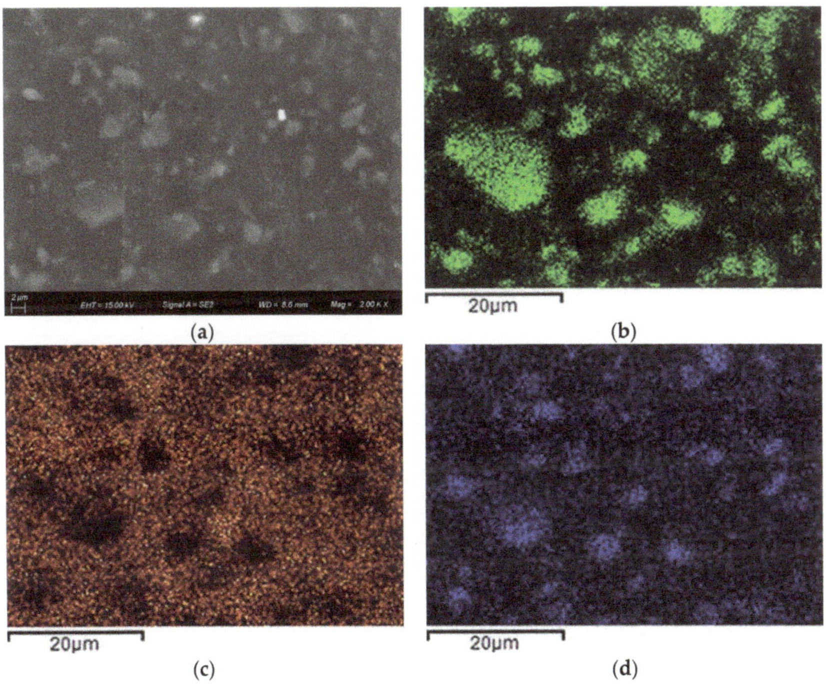

Figure 7. SEM micrographs and elemental mapping of coating films with the addition of ATH. (**a**) SEM image of formulation H-0.75; (**b**) Aluminum map; (**c**) Carbon map; (**d**) Oxygen map.

3.2.3. Effect of ATH on the UV Resistance of the Films

UV light degrades coatings by transferring its energy into the films. Its energy can cause damages by creating heat or by actually breaking chemical bonds in a crosslink. Both the heat and the breakage of bonds can create a loss of physical properties in the coating films, such as color change, gloss change and chalking.

Figure 8 presents the gloss retention (normalized value) at 60° of UV accelerated test against 1000 h of exposure attained for the four resin systems and at different P/Bs. The UV resistance of different resin systems (ATH-free) is in the order of polyester, polyurethane, hybrid and epoxy. For the ATH-free formulations based on epoxy (E) and hybrid (H), the gloss retentions deteriorate very fast and decreased to 90% within 130 h. While PE based powder coatings exhibit the best UV resistance performance, the first 10% decrease of the gloss retention takes place after 690 h of exposure in the UV resistance test. The gloss retention against UV exposure of formulations based on polyester is 5 times over that of epoxy.

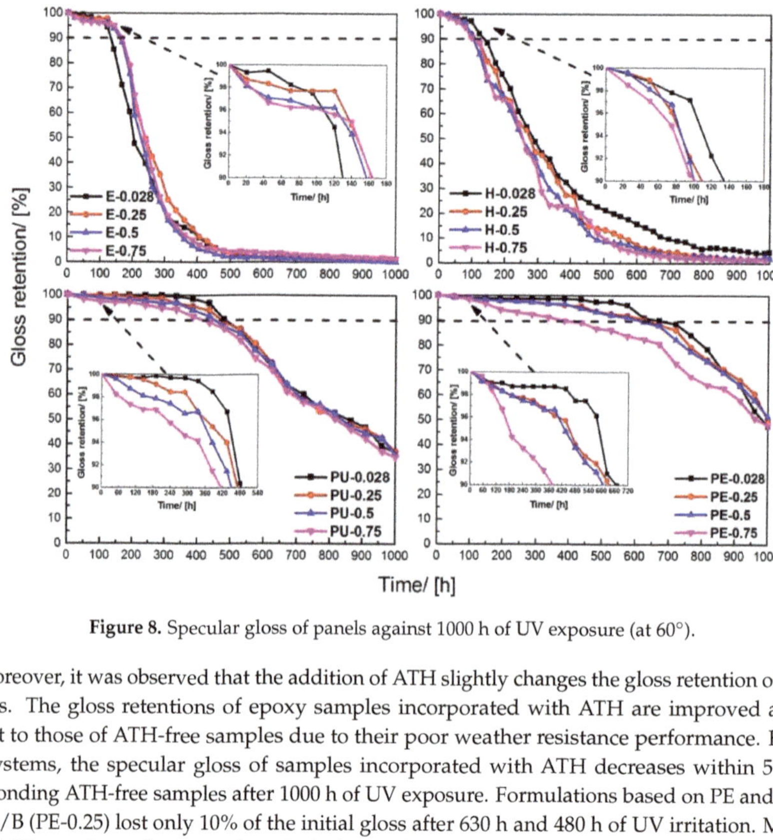

Figure 8. Specular gloss of panels against 1000 h of UV exposure (at 60°).

Moreover, it was observed that the addition of ATH slightly changes the gloss retention of powder coatings. The gloss retentions of epoxy samples incorporated with ATH are improved a little in contrast to those of ATH-free samples due to their poor weather resistance performance. For other resin systems, the specular gloss of samples incorporated with ATH decreases within 5% of the corresponding ATH-free samples after 1000 h of UV exposure. Formulations based on PE and PU with a 0.25 P/B (PE-0.25) lost only 10% of the initial gloss after 630 h and 480 h of UV irritation. Moreover, the PE-0.25 sample retains more than 50% of specular gloss even after exposure of 1000 h in the UV accelerated test chamber. Compared with other formulations, polyester powder coatings achieved the best performance in the UV resistance test. It had the highest gloss retention even at the maximum ATH loading. For example, PE-0.75, the sample panel kept 90% of the initial gloss for 380 h and 50% of gloss for 960 h long. The PE based powder coatings exhibit an outstanding performance when used in exterior conditions.

As to gloss reduction ($Gloss_{t=0}$-$Gloss_{t=1000}$), due to the fact that the addition of ATH significantly decreased the initial specular gloss (Section 3.2.2), the gloss reduction of ATH-free formulations by UV exposure is much more significant than formulations incorporated with ATH.

3.2.4. Effect of ATH on Corrosion Resistance of the Film

Figure 9 shows the results of corrosion resistance tests from the formulations based on the four resin systems with different P/B ratios. The type of polymer resins has an important effect in the corrosion resistance of the paint powders. When no filler is added to the formulations, the rating numbers are grade 8, 9, 7 and 8 for ATH-free formulations based on epoxy, hybrid, polyurethane and polyester, approximately. Moreover, it is also observed that the addition of ATH slightly decreases the corrosion resistance performances of powder coatings. The rating numbers (or mean rust creepage) of formulations based on polyurethane and polyester decrease only 1 grade (within 2 mm) at higher P/Bs ($\geq$0.5), while other formulations based on hybrid and epoxy remain stable in the whole range of P/B. Furthermore, it is important to mention that, although the rating numbers of formulations based on hybrid are grade 9, the length and width of creepages along the scribe show a little deterioration with the increasing of the P/B, as shown in Figure 10.

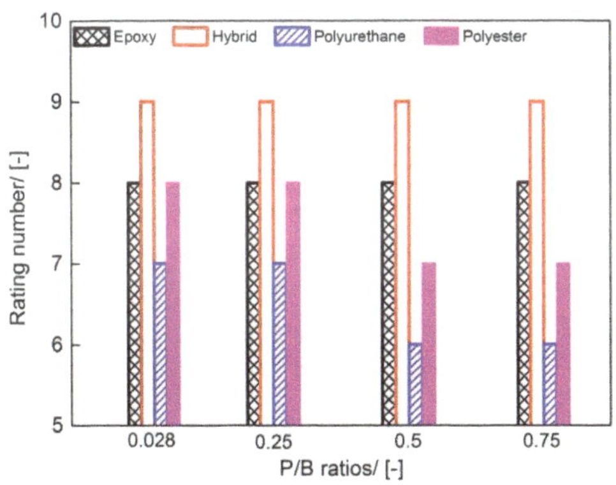

Figure 9. Rating number of failure at scribe of formulations with different resins and P/Bs after 1000 h of salt spray test.

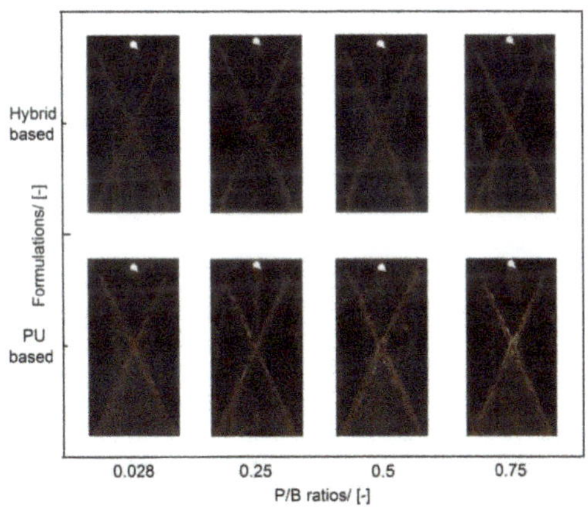

Figure 10. Aspect of the steel panels coated with powder coatings based on different resins and P/Bs after 1000 h salt spray test.

Apparently, the coating film is a good barrier to water and other contaminants when it is continuous. But actually, the film surface before corrosion resistance test has visible voids with the addition of ATH, as shown in Figure 6b. As mentioned before, the film shrinkage with the presence of a high number of ATH agglomerates enables the formation of a micro-rough film surface structure. In corrosion resistance test, when the NaCl solution keeps penetrating the coating film (polymeric matrix) through the voids and reaches the metal surface (metals surface), metal corrosion occurs, which finally leads to the propagation of fissures and the loss of adhesion and delamination of the coating film. This can be attributed to the very low level of binder in the coating at these high P/B ratios (0.5 or 0.75) as well.

4. Conclusions

In this study, ATH, a widely used fire retardant filler in solvent-borne coatings, was incorporated in ultrafine powder coatings based on four widely used resin systems with different P/Bs. Essential performance characterizations on these powders and the coating films made from them were conducted. The effects of ATH on flowabilities of powder paints and physical, mechanical properties, UV and corrosion resistance performances of coating films were carefully investigated.

The increase of AOR of powder paints with addition of ATH indicates slightly flowability reduction, while with the help of fluidization additives, the AOR remains in an acceptable range (<42°) except for those of powders based on PU and PE at their high P/B ratios (0.5 or 0.75), and the powders could be expected to have a good flow behavior during the electrostatic spraying process. It is worth mentioning that a remarkable enhancement is shown in matting effect because of the micro-rough surface formed by the floatation and agglomeration of ATH filler particles. No obvious deterioration appears in mechanical properties of coating films. It shows a significant improvement in pencil scratch hardness. The flexibility properties of films remain stable in the whole range of P/B. While the impact resistance decreases slightly due to the discontinuities in the matrix. A slight decrease can be observed in UV and corrosion resistance. Gloss retention of most formulations drops slightly with the increasing of P/B, while samples exhibit excellent UV resistance performances especially polyester based formulations, even at the maximum ATH loading. The corrosion resistance deteriorates only for PU and PE base formulations at P/Bs over 0.5.

The evaluation results provide valuable information for the application of ATH fillers to powder coatings. This study suggests ATH is a good alternative for powder coating industries to achieve lower gloss finishes, considering its great economic advantage and strong matting effect of powder coatings containing ATH fillers.

Author Contributions: Conceptualization, J.Z. and H.Z. (Hui Zhang); Methodology, D.C.F.; Software, M.S.Y.; Validation, H.Z. (Hui Zhang), H.Z. (Haiping Zhang) and W.L.; Formal Analysis, W.L.; Investigation, W.L. and D.C.F.; Resources, H.Z. (Hui Zhang) and Y.S.; Data Curation, W.L.; Writing—Original Draft Preparation, W.L. and X.Z.; Writing—Review and Editing, H.Z. (Hui Zhang) and H.Z. (Haiping Zhang); Visualization, W.L.; Supervision, H.Z. (Hui Zhang); Project administration, J.Z.; Funding acquisition, H.Z (Hui Zhang).

Funding: This research was funded by the Natural Sciences and Engineering Research Council of Canada (NSERC), Discovery Grant RGPIN-2018-06256.

Conflicts of Interest: The authors declare no conflict of interest.

Notations

D_{50}	Median particle diameter (50 vol.% of the powder smaller than the diameter) [μm]
$\text{Gloss}_{t=0}$	Initial gloss [-]
$\text{Gloss}_{t=x}$	Gloss value at a general time "x" [-]

References

1. Zhu, J.X.; Zhang, H. Ultrafine powder coatings: An innovation. *Powder Coat.* **2005**, *16*, 39–47.

2. Misev, T.A.; Van der Linde, R. Powder coatings technology: New developments at the turn of the century. *Prog. Org. Coat.* **1998**, *34*, 160–168. [CrossRef]
3. Huang, Q.; Zhang, H.; Zhu, J. Flow properties of fine powders in powder coating. *Particuology* **2010**, *8*, 19–27. [CrossRef]
4. Wei, H.; Liang, H.; Zou, J.; Shi, W. UV-curable powder coatings based on dendritic poly(ether-amide). *Prog. Org. Coat.* **2003**, *90*, 287–291. [CrossRef]
5. Stojanović, I.; Šimunović, V.; Alar, V.; Kapor, F. Experimental evaluation of polyester and epoxy–polyester powder coatings in aggressive media. *Coatings* **2018**, *8*, 98. [CrossRef]
6. Zhu, D.; Luo, F.; Xiong, L.; Zhou, W. Preparation and properties of glass coats containing SiCN nano powder as filler. *Mater. Sci. Eng. A* **2006**, *431*, 311–314. [CrossRef]
7. Luo, S.; Zheng, Y.; Li, J.; Ke, W. Effect of curing degree and fillers on slurry erosion behavior of fusion-bonded epoxy powder coatings. *Wear* **2003**, *254*, 292–297. [CrossRef]
8. Tang, Q.; Zhang, Y.M.; Zhang, P.G.; Shi, J.J.; Tian, W.B.; Sun, Z.M. Preparation and properties of thermal insulation coatings with a sodium stearate-modified shell powder as a filler. *Int. J. Miner. Metall. Mater.* **2017**, *24*, 1192–1199. [CrossRef]
9. Leong, C.K.; Chung, D.D.L. Improving the electrical and mechanical behavior of electrically conductive paint by partial replacement of silver by carbon black. *J. Electron. Mater.* **2006**, *35*, 118–122. [CrossRef]
10. Shang, Q.; Hao, S.; Wang, W.; Fu, D.; Ma, T. Preparation and characterization of antistatic coatings with modified $BaTiO_3$ powders as conductive fillers. *J. Adhes. Sci. Technol.* **2013**, *27*, 2642–2652. [CrossRef]
11. Howell, D.M. Powder Coatings. In *the Technology, Formulation and Application of Powder Coatings*; John Wiley and Sons Ltd.: Hoboken, NJ, USA, 2000.
12. Liberto, N. *User's Guide to Powder Coating*, 4th ed.; Society of Manufacturing Engineers: Dearborn, MI, USA, 2003.
13. Valadez-Gonzalez, A.; Cervantes-Uc, J.M.; Veleva, L. Mineral filler influence on the photo-oxidation of high density polyethylene: I. Accelerated UV chamber exposure test. *Polym. Degrad. Stab.* **1999**, *63*, 253–260. [CrossRef]
14. Wei, W.; Guo, J. The investigation of blended conductive coating of graphite/carbon-black. *J. Xi'an Jiaotong Univ.* **1994**, *28*, 98–103. (In Chinese)
15. Wypych, G. *Handbook of Fillers*, 4th ed.; ChemTec Publishing: Toronto, ON, Canada, 2016.
16. Qiu, F.; Wang, M.; Hao, Y.; Guo, S. The effect of talc orientation and transcrystallization on mechanical properties and thermal stability of the polypropylene/talc composites. *Compos. Part A* **2014**, *58*, 7–15. [CrossRef]
17. Shafiq, M.; Yasin, T.; Aftab Rafiq, M. Structural, thermal and antibacterial properties of chitosan/ZnO composites. *Polym. Compos.* **2014**, *35*, 79–85. [CrossRef]
18. Seyedmehdi, S.A.; Zhang, H.; Zhu, J. Superhydrophobic RTV silicone rubber insulator coatings. *Appl. Surf. Sci.* **2014**, *258*, 2972–2976. [CrossRef]
19. Xanthos, M. *Functional fillers for plastics*; Wiley-VCH: Weinheim, Germany, 2005.
20. Cárdenas, M.A.; García-López, D.; Gobernado-Mitre, I.; Merino, J.C.; Pastor, J.M.; Martínez, J.D.D.; Barbeta, J.; Calveras, D. Mechanical and fire retardant properties of EVA/clay/ATH nanocomposites—Effect of particle size and surface treatment of ATH filler. *Polym. Degrad. Stab.* **2008**, *93*, 2032–2037. [CrossRef]
21. Khattab, M.A.; Feteha, F.A.H.; Sadik, W.A.; Abdel-Bary, E.M. Effect of aluminum trihydrate as flame retardant on properties of a thermoplastic rubber nanocomposite. *Fire Mater.* **2017**, *41*, 688–699. [CrossRef]
22. Müller, B.; Poth, U. Paint formulation. In *Coatings Formulation*, 3rd ed.; Vincentz Network: Hanover, Germany, 2017; Part I; pp. 60–65.
23. Cheremisinoff, N.P.; Cheremisinoff, P.N. *Hydrodynamics of Gas-Solids Fluidization*; Gulf Publishing Co.: Houston, UK, 1984.
24. Visser, J. Van der Waals and other cohesive forces affecting powder fluidization. *Powder Technol.* **1989**, *58*, 1–10. [CrossRef]
25. *ASTM D5965-02(2013) Standard Test Methods for Specific Gravity of Coating Powders*; ASTM International: West Conshohocken, PA, USA, 2013.
26. *ASTM D3363-05(2011) Standard Test Method for Film Hardness by Pencil Test*; ASTM International: West Conshohocken, PA, USA, 2011.

27. *ASTM D2794-93(2010) Standard Test Method for Resistance of Organic Coatings to the Effects of Rapid Deformation (Impact)*; ASTM International: West Conshohocken, PA, USA, 2010.
28. *ASTM D522/D522M-13 Standard Test Methods for Mandrel Bend Test of Attached Organic Coatings*; ASTM International: West Conshohocken, PA, USA, 2013.
29. *ASTM D523-14(2018) Standard Test Method for Specular Gloss*; ASTM International: West Conshohocken, PA, USA, 2018.
30. *ASTM B117-11 Standard Practice for Operating Salt Spray (Fog) Apparatus*; ASTM International: West Conshohocken, PA, USA, 2011.
31. *ASTM D1654-08(2016) Standard Test Method for Evaluation of Painted or Coated Specimens Subjected to Corrosive Environments*; ASTM International: West Conshohocken, PA, USA, 2016.
32. Hess, M. The relationship between gloss and surface roughness of paint films. *J. Oil Colour Chem. Assoc.* **1956**, *39*, 185–192.
33. Flectcher, T.E. A simple model to describe relationships between gloss behavior, matting agent concentration and the rheology of matted paints and coatings. *Prog. Org. Coat.* **2001**, *44*, 25–36. [CrossRef]
34. Liu, G.H.; Zhou, B.H.; Li, Y.F.; Qi, T.G.; Li, X.B. Surface properties of superfine alumina trihydrate after surface modification with stearic acid. *Int. J. Miner. Metall. Mater.* **2015**, *22*, 537–542. [CrossRef]

coatings

Review

Powder and High-Solid Coatings as Anticorrosive Solutions for Marine and Offshore Applications? A Review

Krystel Pélissier * and Dominique Thierry

French Corrosion Institute, 220 Rue Pierre Rivoalon, 29200 Brest, France; dominique.thierry@institut-corrosion.fr
* Correspondence: krystel.pelissier@institut-corrosion.fr

Received: 27 August 2020; Accepted: 22 September 2020; Published: 24 September 2020

Abstract: The constant change in coating technology and environmental regulations has led to the development of low-solvent to solvent-free organic coatings, such as powder and high-solid coatings. These two technologies are well developed, but are still not the preferred choice when considering anticorrosive coating for marine and offshore applications. This mostly arise from a lack of perspective in their long-term behavior and from the difficulty in their applications. This review's principal aim is to describe powder and high-solid technologies from their formulation and application to their use in marine and offshore applications while recalling and giving key notions needed when the field of anticorrosive coatings is considered. First, the requirement for coatings to be called anticorrosive will be given alongside with their protection and failure mechanisms. The formulation and application for high-solid and powder coatings will be exposed followed by the description of the type of coating chemistry used in harsh environment. Finally, high-solid and powder coatings behavior in these types of environment will be discussed.

Keywords: low VOC; powder coating; high-solid coating; marine atmosphere; offshore applications; FBE

---

1. Introduction

The high potential of metals, in particular steel, to get oxidized by oxygen and water make them sensible to degradation both in terms of appearance and integrity damages. These degradations due to corrosion can represent a non-negligible cost of the gross annual product of industrial countries [1]. That is why it has always been crucial to find means to protect metals against corrosion. The traditional way to do it, is to apply organic coatings as they can delay the spontaneous corrosion events [2]. Throughout the last decades, both organic and inorganic coatings have been widely applied for the protection of metals against corrosion. This extensive use of coatings has led to a certain maturity of the coating industry. As there are continual changes in technology, mostly due to international and national legislations aiming at reducing the use of volatile organic compounds (VOCs), the coating industry is always reinventing itself. For example, traditional coatings contain a relatively large amounts of organic solvents, thus, reducing VOCs means a significant change in formulation. In order to be able to comply with VOCs reducing policy, the coating industry has designed different types of products such as waterborne coatings, high-solid coatings and powder coatings.

These types of formulation have been largely developed, however, the market for low VOCs coatings is still growing. It is projected to grow by roughly 130% in five years in terms of United States Dollar (USD) [3]. This means the beginning of the end of the long running dominance of traditional solvent-borne coatings in favor to solvent-free coatings. As for traditional solvent-borne coatings, these types of low to free solvent coatings can be used in various applications, from decorative applications (indoor or outdoor) to protection of steel in harsh environments (e.g., marine and offshore environments). Although, high-solid and powder coatings are becoming frequently applied, it may be difficult to substitute completely traditional solvent-borne coatings in harsh environments.

Even though, these types of coating possess the same base chemistry and same protection mechanisms as traditional solvent-borne coatings, their systematic use is still not established and waterborne coatings are preferred. Furthermore, despite continuous development in coating technologies and understanding of anticorrosive coatings degradation, long-term protection of metal from aggressive environments is still an up-to-date subject. This arises mostly from the complexity of the coating-substrate system and the number of factors affecting the performance and service life of anticorrosive coating [2,4–6].

The main objective of this publication is to present powder and high-solid coatings technology in terms of both formulation and application, to describe how they protect steel, how they can be degraded and their behavior when exposed to harsh environments. This review also aims at giving all the necessary key concepts to understand the problematic of corrosion protection by anticorrosive coatings in harsh environments whether it is solvent-borne or low VOCs coatings as their purpose is the same (i.e., protect steel) even though all these notions were already developed in others publications such as the work of Sorensen [1].

2. Protection and Failure Mechanisms

2.1. What Is an Anticorrosive Coating?

Before reaching the heart of this publication subject, it seemed important to describe what is an anticorrosive coating as it is this aspect which will allow them to protect steel from corrosion. It should be noted that as powder and high-solid coatings are similar in terms of chemistry as solvent-borne coatings, first a general description of what is an anticorrosive coating, how they protect the steel and how they fail will be made.

An anticorrosive coating is, typically, applied on a metallic surface (in this case, steel) to protect it by stopping or at least slowing down corrosion that can spontaneously occurs when exposed to any corrosive environment. They are constituted by five main groups of components; the binder, pigments, fillers additives and carrier. They can be formulated from a wide variety of chemicals, materials or combination of different chemicals. Additives are secondary products included in small quantities to improve selected coating properties or to solve technical problems.

Among the components of anticorrosive coatings, the carrier is of importance even though it will not have a crucial role in the mechanisms of protection offer by the coatings. It is the vehicle in the uncured paint that carries all elements and exists only in the uncured state. In addition to dissolve or disperse the constituents, the carrier can reduce the viscosity of the coating leading to an easier application and enable the application by spraying or dipping. The carrier can be an organic solvent, water or gaseous depending on the type of coatings. In its liquid form, the carrier will be mostly responsible for the coating wetting of the surface and thus control the homogeneity of the coating film. If the wetting is insufficient, it can lead to holes in the film and unprotected areas.

The binder is what brings the physical structure of the polymer and permits to support and contains the pigments and additives. It binds itself to these components and to the metal surface. In addition to bring structure, it ensures the most important properties of the coating (adhesion to the metal, cohesion within the coating, high mechanical strength and low permeability). It is, thus, evident that the right binder choice is crucial when designing an anticorrosive coating as its properties will depend on its chemical composition and curing. Very often, coating performance will be linked to the advantages and disadvantages of its binder family.

It exits three main types of pigments: barrier, sacrificial and inhibitive pigments. They differ in their mechanisms of protection and also in their quantity in the coating as depicted on Figure 1. Indeed, sacrificial coatings (typically zinc-rich coatings) need to have a large quantity of pigment in them to allow the flow of electric current coming from the electrical contact between the particle and the substrate to be protected. Inhibitive pigments which rely on a dissolution mechanism and formation of soluble salts at the substrate/paint interface needs to be in sufficient quantity whereas pigments in barrier coatings are generally chemically inert.

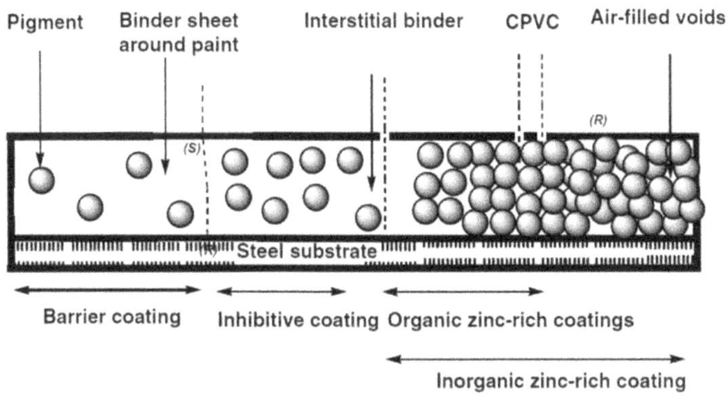

Figure 1. Pigment volume concentration (PVC) as a function of the type of anticorrosive coating. CPVC stands for critical pigment volume concentration. Reproduced with permission from [7]. Copyright 2008 Elsevier.

Among the five groups, binder and pigments are the most important elements when considering the anticorrosive performance. Whereas the binder will give the overall properties of anticorrosive coatings, pigments will be crucial when considering the protection mechanisms.

Now that the principal components of an anticorrosive coating have been reviewed, let us see how they can provide protection to the steel. First, an anticorrosive coating will act as a barrier between the potentially corrosive environment and the steel. To be able to understand how these coatings can protect steel, it is important to briefly recall the definition of corrosion and the corrosion process. Corrosion can be defined as a physicochemical interaction between the environment and a material that results in changes in the property of the metal (metal consumption/dissolution). It is a process driven by a potential difference between cathode and anode sites. The anode and cathode sites can be next to one another or separated from one another. The anode site is at the metal surface's less noble site or site with a higher surface energy. Corrosion requires several elements to happen such as water, oxygen or another reducible species, a dissolution process at the anode, a cathode site and an electrolyte path between the anode and the cathode. As any of these elements can control the rate of corrosion, suppressing one or several of them is the main goal to achieve when designing an organic coating.

Consequently, the main protection mechanisms displayed by organic coating are:

- Creating an effective barrier against the corrosion reactants (water, oxygen and ions);
- Creating a path of extremely high electrical resistance, thus inhibiting anode-cathode reactions;
- Passivating the metal surface with soluble pigments;
- Providing an alternative anode for the dissolution process.

Based on these mechanisms and according to Bierwagen [2], there are three main modes of corrosion protection that anticorrosive coating can provide: (i) barrier protection, (ii) cathodic/sacrificial protection and (iii) inhibitive/passive protection (coating formulated with inhibitive pigments/or additives).

It exists a large body of literature on the protection mechanisms of anticorrosive coatings as these subjects has been extensively studied [2,4,8,9]. The aim of this section is not to detail the protection mechanisms of anticorrosive coatings but to give a critical review of them.

2.2. Protection Mechanisms

As said, traditionally the protection mechanisms described in the literature are the barrier protection, the sacrificial protection and the inhibitive/passive protection. The barrier protection depends at the same time on the coating properties and the presence of pigments whereas the sacrificial and inhibitive/passive protection mostly depend on the presence of pigments. This section will focus on the role of pigments as the case of the polymer properties due to binder choice will be detailed later in this review.

2.2.1. Barrier Protection

Barrier protection, as its name implies it, relies on the creation of a barrier which will reduce the transport of species needed in the corrosion process, i.e., water, oxygen but also chemical species such as H_2O, O_2 and electrolyte ions (Cl^-, SO_4^{2-}, Na^+, K^+, NH_4^+ and Ca^{2+}). It is a crucial protection and was at first stated as the main protection mechanism offered by coating but the permeability of coating to water and oxygen discredited it [10]. It was shown that the extent of water and oxygen transported through the coating was at least an order of magnitude greater or comparable to the amount needed to support corrosion [10,11].

Concerning the impediment of the transport of electrolyte ions, it is the resistance to flow of charge which is of importance [2]. When transported to the interface, electrolyte ions can react with metallic species and formed soluble complexes which will diffuse away from the corrosion sites and stimulate the corrosion cycle [12–16]. That is why, allowing the diffusion of water without ions is essential. For Mayne, this coating property (ionic resistance) is the variable controlling the degree of corrosion but the recent work of Sykes et al. [17] showed that an undisputable correlation between conductivity and corrosion protection was not possible. They tried to measure the high resistance pathway by electrochemical impedance spectroscopy (EIS) using bi-electrodes and showed that it was not measurable. They also carried out test at elevated temperature and demonstrated that the coating resistance measured could not control the corrosion process.

Protection offered by barrier coating is known to be greatly dependent on a number of parameters such as the polymer properties, the polymer layer structure (crosslinking density) and thickness, the type of corrosive environment, its interaction with the substrate, the presence of a pre-treatment layer, pigment volume concentration, type of pigments/fillers and particle shape of pigments/fillers (see Figure 2) [1,5]. It should be noted that barrier coating may be used as primer, intermediate or topcoat.

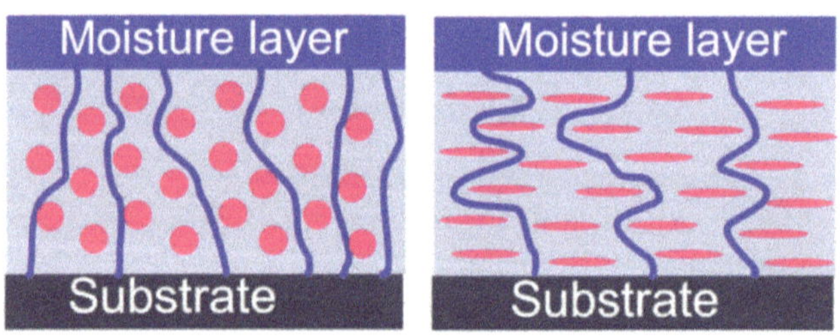

Figure 2. Barrier properties of shaped pigments: spherical pigments allow easy percolation of water while lamellar pigments create a tortuous pathway for diffusion. Reproduced with permission from [5]. Copyright 2017 Elsevier.

Concerning the pigments used in barrier coating, their efficiency to disturb water and ions transport through the coating will depend on their nature and shape as illustrated on Figure 2.

Spherical pigments will allow aggressive species to migrate almost straight through the coatings whereas lamellar pigments will provide a tortuous path for the aggressive species. Pigments used in barrier coatings are inert and insoluble in water. Another parameter to take into account for lamellar pigments is the build-up of layers in the dry film as the more layers there are, the more protection they offer. This effect is controlled by the shape and size of the lamellas and in particular the diameter-to-thickness ratio [1,18]. The two main groups of barrier pigments are mineral-based materials mineral-based materials (mica, micaceous iron oxide (MIO) and glass flakes) and metallic flakes (aluminum, zinc, stainless steel, nickel or cupronickel). For more information about the different types of barrier pigments please refer to [1,18–21].

2.2.2. Sacrificial Protection

In the case of sacrificial coating, the protection is offered by galvanic effect. The substrate will be protected by a metal or an alloy that is electrochemically more active than the material to be protected. Indeed, the protective metal or alloy will be dissolved preferentially and leave the steel unreacted.

In the case of a coating, the cathodic protection is achieved by pigmentation. The major pigments used to achieve sacrificial protection are zinc pigments. For the coating to be able to offer cathodic protection, the content of sacrificial pigments must be high enough to ensure an electrical contact between them (just below the critical PVC). Thus, as long as the conductivity in the paint is preserved (enough pigments remaining to act as an anode), the steel will be protected. In the case of zinc rich paints (ZRP), the weight of zinc must be higher than 80% in the dry film. Another important parameter to control is the shape and the size of the zinc particles, although the size of the zinc particles will be more important.

This type of protection is active when the coating is damaged. It will lead to the consumption of the zinc particles and corrosion products will be produced as illustrated on Figure 3. As it can be seen on this figure, in addition to the cathodic protection, the metallic powder can also offer barrier protection due to the build-up of insoluble corrosion products in damage areas. As they precipitate, these products fill the pore of the coating, thus, reducing the permeability of the film [22–24].

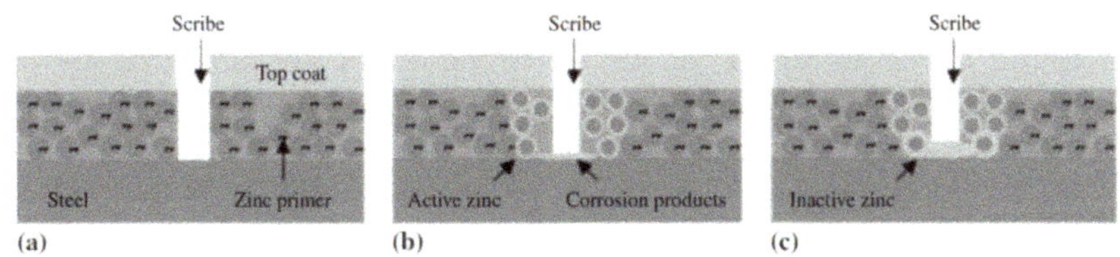

Figure 3. Sketch of the working mechanisms of a zinc-rich coating system. (**a**) Damage is introduced down to the steel surface; (**b**) zinc around the score is still active while zinc-iron corrosion products are being built-up; (**c**) the zinc is not galvanically active any longer but the coating provides barrier and inhibitive protection. Reproduced with permission from [1]. Copyright 2009 Springer Nature.

Sacrificial coatings are only applied as primers as they are effective only if the coating is in direct contact with the steel due to the requirement of electrical contact between the steel and the sacrificial coating. Since these types of coatings contain a low content of binder and other components, their other properties (mechanical properties, adhesive and cohesive strength properties, impact resistance) are significantly reduced [1].

2.2.3. Inhibitive/Passive Protection

Similar to sacrificial protection, inhibitive/passive protection relies on action of pigments when the surface to be protected will be in contact with moisture. Their mechanism of action is different than sacrificial pigments even though the first step is the dissolution of inhibitive pigments. In general, such pigments are chosen for their transformation in insoluble metallic complexes when reacting with the steel surface after their partial dissolution due to water penetration in the coating [2,25]. As the insoluble metallic salts build-up, they passivate the steel by impeding the transport of aggressive species acting as a barrier.

Inhibitive pigments are inorganic salts which are slightly water soluble. The most common inhibitive pigment is based on phosphate chemistry and mainly zinc phosphate is used. Zinc phosphates are widely used because of their low solubility and activity which make them compatible with a vast type of binders however their performance as corrosion inhibitor is contradictory but it is admitted that they give good performance in industrial environments [26]. Other types of phosphates can also be found, such as aluminum, calcium, barium, iron, cadmium or magnesium phosphates. Inhibitive pigments can have two types of inhibitive actions; either an anodic inhibition by maintaining a passive layer or a cathodic inhibition by reducing the corrosion rate by restricting oxygen diffusion to the surface [1]. They are several mechanisms by which zinc phosphates protect the steel and they were largely described in the literature [18,26–31]. It exists a vast variety of type of zinc phosphates which were detailed with great care by Knudsen and Forsgren [4]. The mechanism of protection of other zinc phosphates pigments is interpreted similarly as that of zinc phosphate [27,32,33]. Among the other types of inhibitive pigments, ferrites, calcium exchanged silica, barium metaborate, molybdates and silicates can be noted [18,27,34,35].

The design of inhibitive coating is delicate because their efficiency relies on antagonistic requirements as they need water to be dissolved and act as a barrier after dissolution. Then, a balance between the barrier properties and the effectiveness of the inhibitor must be achieved. The degree of protection from an inhibitive/passive layer will then depends on the nature of inhibitive pigments and also on the pigment solubility in the organic coating. As sacrificial coatings, these types of coating are essentially applied as primers because they need to react with the steel surface to be protective.

2.2.4. Adhesion: A Protective Mechanism or a Pre-Requisite for Corrosion Protection?

According to Funke [36], "Adhesion is the most important property of organic coating because if it fails, all other coating properties become worthless" but he also defined adhesion as a protective mechanism along with the electrochemical mechanism (protection by pigments) and the physicochemical mechanism (barrier protection) [37]. It is true that in most of publications about corrosion protection, the three mechanisms of protection proposed are the ones already described previously and adhesion is not mention alongside them.

Adhesion can be described as the interatomic and intermolecular forces at the interface of two surfaces but in reality, it is very complex phenomenon implicating a multidisciplinary knowledge. Numerous studies have been performed on adhesion to describe it, to explain how it works and how it can be predicted as it is a key aspect in the protection offered by coating [38–44]. Several theories exist to explain adhesion but the aim of this paragraph is not to review them and detailed information can be found elsewhere [38–44]. Very strong adhesion can help suppress corrosion by retarding the formation of corrosion products, hydrogen evolution and water build-up under the coating. Basically, the role of adhesion is to create the necessary conditions so that corrosion protection mechanisms can work. Indeed, without an intimate contact with the metallic surface at the atomic level, organic coating cannot act as a passivation layer, create a path of extremely high electrical resistance at the metallic surface or prevent the contact with electrolyte or corrosive species. However, adhesion is only a "necessary but not sufficient" condition for good corrosion protection. Good adhesion alone is not enough to ensure protection as it will not necessarily imply enhanced corrosion protection [45]. It is particularly the case when considering dry adhesion which is the initial strength of the coating-substrate bond. Funke himself said that when considering protection against corrosion, dry adhesion was less decisive than wet adhesion [36].

Wet adhesion can be defined as the adhesion of an organic coating to a support upon exposure to liquid water or high humidity [36]. Wet adhesion is a coating property and not a failure mechanism. The same polar groups on the binder molecules that create good dry adhesion can decrease the water resistance at the coating-metal interface and thus the wet adhesion. Based on their literature review,

Leidheiser and Funke [46] proposed two disbonding mechanisms; the chemical disbondment where the water molecules can disrupt the interactions between the polymer and the metal oxide and the loss of adhesion due to a mechanical disbondment originating from accumulation of water at the interface, see Figure 4.

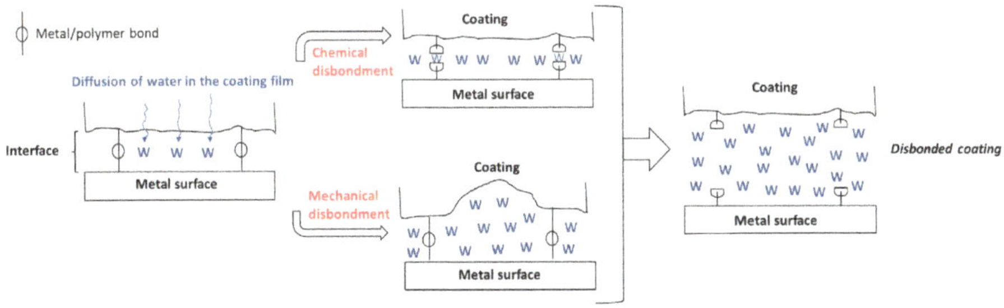

Figure 4. Schematic representation of the mechanisms of adhesion reduction and water disbondment [46].

In a general point of view, the adhesion mechanisms are controlled by the physicochemical properties of the polymer (chemical composition and spatial distribution of the components, the rheology of the polymer, the glass transition temperature (Tg) and the mechanical properties) and the metal (nature of the reactive species and their spatial distribution on the metallic surface, roughness). It is thus difficult to conclude about the role of adhesion as it could be considered as an essential condition for corrosion protection to work and also as a protection. Nevertheless, this coating property is of upmost importance and need to be carefully considered.

2.3. *Other Important Aspects Related to Anticorrosive Performance*

It is important to consider other coating aspects such as surface preparation or internal stresses in the coating in addition to the protection mechanisms as both can impact coating performance when stressed and need to be taken account in the design and application of anticorrosive coatings.

2.3.1. Surface Preparation

As stated above, adhesion between the organic coating and the steel surface is a pre-requisite for the anticorrosive coating to actually protect the steel. It is, thus, vital to correctly prepare the surface before the coating application as presence of contaminants (dust, grease, oils, salts) on the surface of the steel will act as a weak boundary layer. This weak boundary layer could lead to coating failure or can promote the appearance of blistering (see Section 2.4.2 for more details) [47–51]. Surface preparation is also important as it can enhance mechanical or chemical anchoring of the coating and several standards can be used as guide for surface preparation [52,53]. Le Calvé et al. [54,55] indicated that the surface cleanliness level of steel was a key parameter in the coating durability. It was shown that coated steel surface with low roughness (machined, Ra 4 µm) presented extremely good performance as long as high level of surface cleanliness was ensured (such as on hydroblasted steel).

The main components of surface preparation are surface cleanliness and anchor profile of the surface. Pretreatment of steel before painting consists in washing the steel with an alkali surfactant, rinsing it with clean water and then removing the mill scale with abrasive blasting. The two first steps provide cleanliness by removing oil, greases or others contaminants which might have accumulated during the period between the production the steel and its painting. This step is achieved either by abrasive blasting (dry or wet abrasive blasting) or with water as propellant and abrasive [52,56]. The last step produces also cleanliness but at the same time roughness, thus, better mechanical anchoring for the anticorrosive coating. The effects of different types of cleaning and pretreatment of metallic substrates, as well as application procedures, has been thoroughly described elsewhere [57,58].

2.3.2. Internal Stresses

The existence of internal stress in the coating, which develops due an inability of the coating to shrink, may add further to the complexity of the coating system. Internal stress in coatings can significantly affect the durability of anticorrosive coatings by resulting in loss of adhesion, cracking or cohesive failure [59].

The build-up of internal stress can come from several contributions [60,61]:

- The film formation: solvent evaporation, volume change associated with the curing reaction, etc.;
- Thermal stress induced by differences in the coefficient of thermal expansion of the coating and the substrate at ambient temperature;
- Stress associated with the degradation of the coating polymer;
- Stress induced by the exposition of the system to humidity and temperature.

The build-up of the internal stress due the exposition to a variation of temperature or relative humidity mostly comes from the fact that the adhesion between the substrate and the organic coating prevent stress relaxation by dimensional changes (film constraining by adhesion) due to different thermal and hygroscopic coefficient between the coating and the substrate [62]. A schematic representation of the different stress which can occur in organic coating is reported on the Figure 5 [63]:

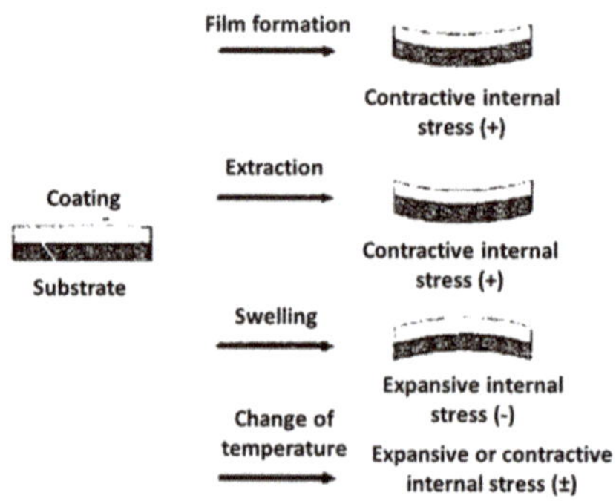

Figure 5. Internal stresses in organic coatings. Reproduced with permission from [63]. Copyright 1996 Elsevier.

In this section, the different protection mechanisms and the requirements for these protection mechanisms to happen were described. It was important to resume them as they will help understanding how anticorrosive coatings work.

2.4. *Mechanisms of Degradation*

In addition to understand the protection mechanism involved in anticorrosive coatings, it is important to understand the possible failure mechanism which can be encountered as a function of the stresses considered. It exists two main types of degradation, the one involving defects (pinholes or mechanical damages) in the coating which can lead to corrosion and the one involving degradation of the coating itself (cosmetic defects or weathering of paint). Both types of degradation are linked to the coating properties such as the resistance to mechanical damage or the formulation of the coating which can induce defects in the paint film or make it more sensitive to weathering.

2.4.1. Loss of Protection Due to Corrosion Initiated at Defects

Coating failure normally initiates at local defects, which can be the result of application errors, chemical heterogeneities in the coating or other defects, such as bubbles, under film deposits or mechanical damages. The corrosion of steel at a defect can lead to cathodic delamination of the surrounding coating or be the result of anodic undermining or filiform corrosion [1,4,20,48,64,65]. For submerged steel, cathodic delamination will be the predominant degradation mechanism whereas in atmospheric weathering conditions alternative mechanisms of de-adhesion can be expected [64]. Another minor mechanism but that can be important is a mechanical delamination due to the accumulation of corrosion products which can expose fresh steel. Which mechanism dominates will depend on the type of coating, type of substrate and exposure conditions. A lot of works has been performed to describe and explain how these mechanisms initiate and propagate and the aim of this section is only to recall the most pertinent information needed for the understanding of these loss of protection.

Cathodic Delamination

Cathodic disbonding is the main mechanism of failure for anticorrosive coatings for submerged and buried steel. However, it can also play a role in degradation mechanism observed on atmospherically exposed steel when the electrolyte is thick enough to induce this type of mechanism or when the adhesion between the paint and the metallic surface is poor. The process of cathodic disbonding starts at coating defects (pinhole), mechanical damages, blister, where the steel is exposed to the electrolyte. Cathodic disbonding is a process causing the organic coating to lose adhesion to the substrate due to the cathodic oxygen reduction:

$$O_2 + 2H_2O + 4e^- = 4OH^-$$

Disbonding from damage is the result of bonds breaking at the coating-metal interface coming from the alkalinity of the cathodic products [1,4,19,63]. They are a result of electrochemical reactions occurring at the coating-steel interface, where solid iron is oxidized to ferrous ions and oxygen is reduced to hydroxyl ions at the metal-paint interface, see Figure 6 [64]. In addition to the production of OH^-, cations should also be available to maintain electroneutrality. The mechanism of loss of adhesion caused by an alkaline environment can be divided into three groups according to the literature; reduction of the oxide layer, chemical degradation of the coating or interfacial failure. All those three mechanisms were experimentally observed and proposed by different authors [1,66,67].

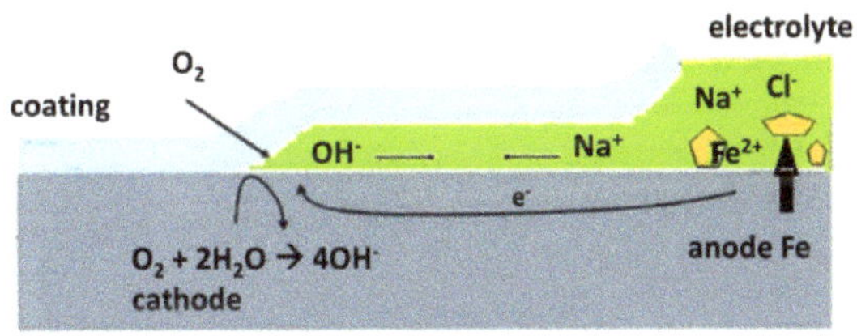

Figure 6. Schematic for corrosion-driven coating de-adhesion in the case of cathodic delamination [64].

The extent of cathodic delamination, from a coating point of view, will be dependent of the ability of the interface to sustain an alkaline pH and the ability of the reactive species to diffuse through the coating and along the coating-substrate interface [19,20]. Other parameters, such as the potential of the coated steel (in case of cathodically protected steel), the type of electrolyte and its concentration, the oxygen concentration, the coating thickness will also influence the rate of cathodic delamination [1,4,68,69].

Anodic Undermining

Anodic undermining usually takes place in humid air conditions. In the case of cathodic delamination, the anodic reaction is concentrated in the defect filled by the electrolyte and compensating cathodic process takes place under the coating. Oxygen diffusion through the thick electrolyte layer is the limiting factor and to remove this limitation, the cathodic front spreads from the defect underneath the polymeric coating. In humid conditions (thin electrolyte), the defect area is well aerated and works mostly as a cathode where oxygen reduction reaction can take place, while anodic reaction occurred underneath the coating and spread from the defect, Figure 7 [1,4,64,70,71]. Now, when the anodic reaction is taking place under the coating, chloride ions (Cl^-) will migrate to balance the Fe^{2+} ions that are formed by the anodic reaction.

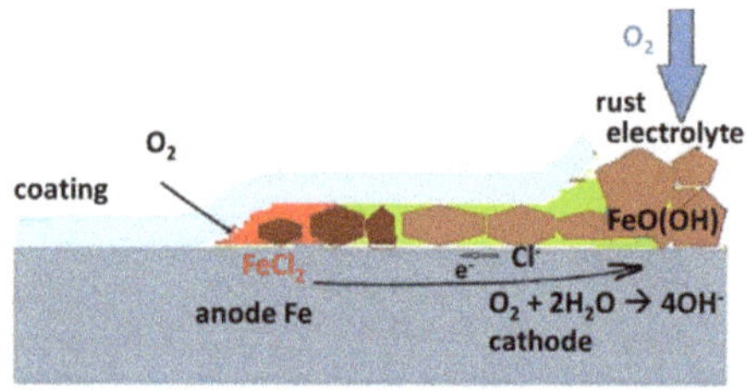

Figure 7. Schematic for corrosion-driven coating de-adhesion in the case of anodic undermining [64].

Hydrolysis of the iron chloride under the coating may result in acidification of the electrolyte. Due to acidification of the electrolyte, hydrogen evolution may start, which will increase the corrosion rate further [4].

Filiform Corrosion

Filiform corrosion is threadlike corrosion which attacks under the coating film, spreading from a coating damage. Although filiform corrosion is usually observed on organic-coated aluminum parts, it has also been observed on magnesium and cold-rolled steel. Filiform corrosion occurs in humid atmosphere. The mechanism of filiform corrosion has been discussed in number of publications such as reference [1,4,48,65,72].

The conditions for initiation and propagation of filiform corrosion are the followings; presence of defects in the coating (mechanical damages, pores, insufficient coverage of edges, air bubbles, salt crystals or dust particles), high humidity air (65–95%R.H), sufficient water permeability of the paint film and stimulation by artificial or natural impurities (sulphur dioxide or chloride) [65].

The filaments consist of an active corroding head followed by an inactive tail filled with porous corrosion products. Oxygen is consumed at the active corroding head which will become deaerated. This results in differential aeration because oxygen and water are supplied to the filament head by diffusion through the porous tail [73,74].

2.4.2. Blistering

Blistering is one of the most important defects which should be avoided when considering the cosmetic aspect of corrosion protection by organic coatings. Blistering is showing a local loss of adhesion between the substrate and the anticorrosive coating. It will impact the appearance and also the performance of the coating. According to Greenfield et al. [47], three types of blistering can be quoted; osmotic, cathodic and anodic blistering. Whereas osmotic blistering is linked to the presence of contaminants at the surface and thus linked to proper surface preparation, cathodic blistering is linked to the presence of defects.

Osmotic blistering mostly originates from the presence of contaminants such as chlorides, sulphates, dust or solvent entrapment [48]. Foreign ions can establish local osmotic cells in the presence of microscopic amounts of water and the water transport is driven by the difference in chemical potential between the outside surface of the film and the inside of the blisters, see Figure 8 [48–50]. The key factors in osmotic blistering are the adhesion at the interface and the metal surface contamination. Blister initiation does not occur when the surface is relatively clean, or where water does not penetrate all the way to the substrate surface [49].

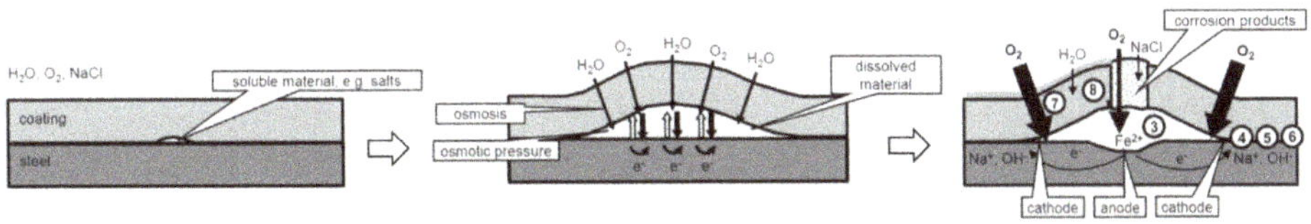

Figure 8. Model for the degradation of an organic coating on steel in a neutral NaCl solution initiated by osmosis. Reproduced with permission from [75]. Copyright 2008 Elsevier.

Cathodic (alkaline) blistering occurs at sites damaged by mechanical action, at a pinhole, at an incomplete or very thin coverage section, or at an area of low ionic resistance or may be inherent fault of the coating (pore). Basically, the mechanism of cathodic blistering is close to the mechanism of cathodic disbonding as a loss of adherence is caused by the generation of a high concentration of OH^- ions in the cathodic zone [47,76]. It is also considered that the cathodic blistering is primarily not due to the loss of adhesion but to the overpressure at the metal interface because of the formation of highly water-soluble corrosion products due to the migration of Na^+ ions at the interface due to the cathodic reaction [76]. The presence of this water-soluble corrosion products could lead to the formation of an osmotic cell leading to the formation of a blister through the same mechanism as osmotic blistering [76]. A model for the degradation of organic coatings with no apparent defect on steel in a neutral electrolyte is represented on Figure 9. In this model, the cations are believed to diffuse through the coating as a result of an attack by water in low molecular weight or low crosslinked regions, followed by the interconnection of these regions.

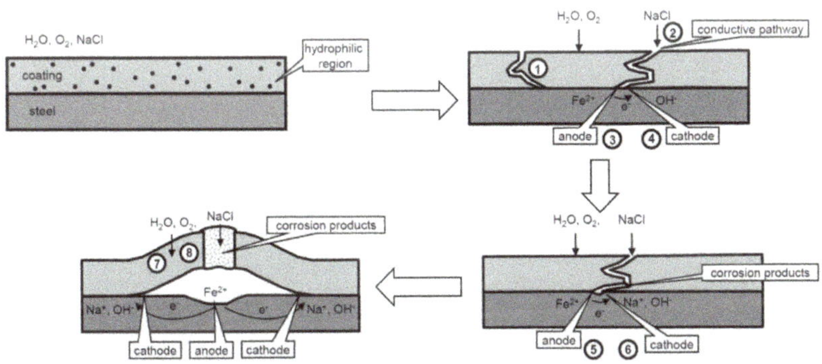

Figure 9. Model for the degradation of an initially intact organic coating on steel in a neutral NaCl solution. Reproduced with permission from [75]. Copyright 2008 Elsevier.

Anodic blistering is much less observed than cathodic blistering. According to Funke [77], on steel, anodic blistering could be the result of a differential aeration process. Polarization could be the result of a different access of the oxygen on a steel surface. The oxygen-poor center of the blister would become anodic while the steel around would be cathodic. This mechanism is illustrated in Figure 10.

Whatever the types of blistering considered it is believed that blistering can be formed by two main mechanisms: through the production of corrosion products under the organic coating or through fully mechanical effect of osmotic pressure. In both cases, blister growth is promoted by the presence of water at metal/polymer interface and interrelated to the loss of adhesion between the organic coating and the metallic surface.

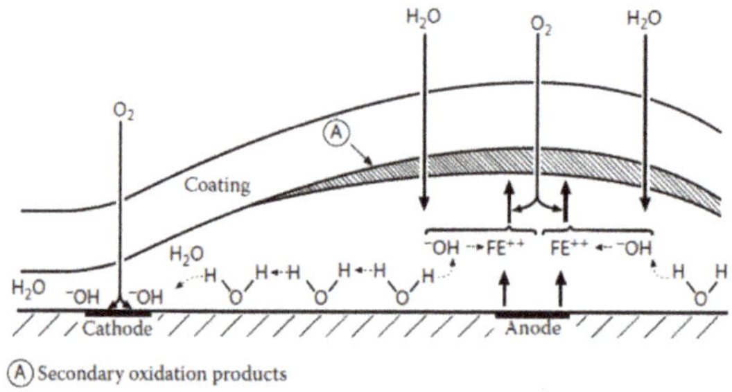

Figure 10. Mechanism of anodic blistering. Reproduced with permission from [77]. Copyright 1985 American Chemical Society.

2.4.3. Loss of Protection Due to the Weathering of the Paint

Even though the degradation of the paint by corrosion and blistering are the most common defects when looking at the failure mechanisms of organic coatings, it is also important to take into account the weathering of the organic coating. Indeed, this weathering can affect the coating properties and thus changes water or electrolyte transport through the coating, i.e., impacting the barrier property of coatings. The aging of coating is inexorable when exposed to weather even for the best coatings, properly applied to compatible substrates. Weathering is a joint action of ultraviolet radiation, heat, oxygen, humidity and chemical species. They can act separately or interact to lead to coating degradation.

The effect of UV from sunlight is usually associated with aesthetic changes such as yellowing, color change, loss of color or chalking. However, the most critical damages are due to chemical breakdown and thus worsened mechanical properties. The range of potential effect of UV includes embrittlement, increased hardness, increased internal stresses, generation of polar groups at the coating surface (increased surface wettability and hydrophilicity), changes in solubility and crosslinking density. The action of UV light is inhomogeneous within the coating thickness as the wavelength range of the UV will have an impact on its penetration depth in the coating [78].

The effect of moisture can come from several sources including water vapor in the atmosphere, rain and condensation due to the day-night cycle. As already said coating are permeable to water or water vapor and it can diffuse through pore or microcracks. This water can be randomly distributed or aggregate into clusters, can create a water interstice between the binder and pigments particles, can exist in voids in the paint film and can accumulate at the metal-coating interface. The uptake of water can affect the coating in several ways such as chemical breakdown, weathering interactions, hygroscopic stress and blistering/adhesion loss.

Temperature can have an impact on the balance of stresses as it can induce additional crosslinking or stresses by difference between thermal coefficient of expansion between the polymer and the metallic surface [60,61,79]. It can also impact the coating mechanical properties as they are closely related to the glass transition temperature (Tg) of the polymer. Diffusion process through the coating can be enhanced by temperature rising as it will increase Brownian motion mode [4].

The chemical degradation is related to the breakdown in the coating film due to chemical contaminants in the atmosphere. When thinking about atmospheric contaminants one will automatically think about their effect at defects, however, these contaminants can also degrade the coating by combining with condensation and make it more vulnerable to UV, moisture or temperature action.

Obviously, the extent of the coating reaction to these stresses will depend on its chemical nature and thus each coating type will react differently to changes in one or more of these weathering stresses.

3. What Are High-Solid and Powder Coatings?

In the previous section, the mechanism by which anticorrosive coatings protect the steel and how they lose their protective properties were exposed. It was shown that it was intrinsically linked to the coating property and how it was applied. High-solid and powder coatings possess the same base chemistry as traditional solvent-borne coatings, however, to achieve a low VOCs formulation, change in fundamental parameters such as the molecular weight of the binder needs to be achieved. This section aims to describe high-solid and powder coatings technology to help understand how it will influence their performance when exposed in harsh environment.

3.1. Why Formulate Low VOCs Coatings

Historically, solvents were used as carrier however their major drawback despite their well-known performance were their hazardous nature for environment as well as human health. Their evaporation at normal temperature and pressure also lead to problematic issues from a safety point of view due to fire or explosion hazards. It was all these problems which generated a reducing VOC policy and the development of environment-friendly solutions.

Volatile organic compounds are emitted as gases from certain solids or liquids and in the scope of this review from the evaporation of organic solvent. VOCs include a variety of chemicals, some of which may have short- and long-term adverse health effects. Nowadays, as companies must control their VOCs, international and national legislation aimed at reducing the use of volatile organic compounds. This trend, led the coating industry to develop environment-friendly solutions such as waterborne coatings, low solvent coatings or solvent-free coatings, see Table 1. Another minor type of environment-friendly coating is radiation curable coatings.

Table 1. Environment-friendly alternatives. Reproduced with permission from [80]. Copyright 2008 Elsevier.

Nomenclature	% Solid Content	Advantages	Disadvantages
High-solid coatings	60–80%	- Higher thickness per coat - Low VOC	- High viscosities - Slight difficulty in brush application
Powder coatings	100%	- Very high thickness per coat	- Costly equipment - Airless spray technique used for coating
Waterborne coatings	Same as solvent-borne coatings	- Low VOC emission - Low viscosity - Reduced toxicity - No odour and non-flammable - Conventional application processes - Suitable for thin film application	- Resins are costly compared to conventional coatings - Tendency to form foam - Requires longer drying times or increased oven temperatures

Generally, whatever the type of coating desired (solvent-borne or solventless), the same binder, curing agent, additives or filler can be found. It is particularly true for high-solid coatings. For powder coatings, the same base chemistry can be found but due to the process of fabrication and application, some specific additives or curing agent can be found.

3.2. High Solid Coating

3.2.1. Generalities

High solid coatings are a group of coatings that highly resemble to the solvent-borne coatings but contain less than 30% solvent. By definition, a high solid coating typically contains more than 60% solids by weight or 80% solids by volume [81,82]. This technology has been accepted by the automotive, aerospace and marine industries for use in a variety of topcoat formulations, such as two-component polyurethanes. High solid coatings use less paint to achieve the same thickness obtained with conventional formulations.

High solid coatings can present several disadvantages among the following can be quoted; higher viscosity, sensitivity to inadequate substrate cleaning, shorter pot life, flocculation of pigment, narrower cure window and sagging.

3.2.2. Parameters Influencing Film Formation

As stressed in the previous section, coating formulation and by extension coating properties as well as the quality of the coating film will have a significant impact on the protective abilities of coatings. That is why it is important to review parameters influencing film formation as it can be different than for traditional anticorrosive coatings.

Viscosity

The same base chemistry as solvent-borne coatings is found for high solid coatings but differs mainly on the molecular weight of the resin [81,83–85]. In contrary to solvent-borne coatings which utilized polymers, high solid coatings are formulated with oligomers with lower molecular weight. The choice of low molecular weight arises because of viscosity issues [86]. As the quantity of solids is higher, the viscosity of high solid coatings is high and can generate difficulty in the application of the paint during the levelling of the film. It exists is a linear relationship between the log of the viscosity and the weight percent of the polymer solids. The increase in viscosity with increasing solids content is much lower with oligomers than with higher molecular weight polymers [81,83].

Molecular weight reduction causes an increase in free volume because chain ends can undergo segmental motion relatively easily and as a consequence decrease the viscosity of the resin which depends on free volume availability [80]. The drawback of this approach is that to obtain adequate film properties, a large increase in Tg will be required during curing. To obtain the necessary properties, the content of functional group needs to be increased to promote molecular weight build-up and crosslinking during curing. As increasing the content of functional groups will also increase the viscosity of the coating, an optimum balance of molecular weight and functional groups content for each high solid system must be found [81,84]. In addition, as the possible presence of single functional group oligomer can lead to the formation of dangling ends which will act as plasticizers, it is thus critical to control the functionality of the oligomers and to have at least two functional groups by oligomer.

Sensitivity to Inadequate Substrate Cleaning—Surface Tension Issue

The increase of functional groups content in high solid coatings can lead to surface tension issues and film defects, such as cratering or crawling [81]. Crawling occurs when a coating of relatively high surface tension is applied to a substrate of relatively low surface free energy. Due to their formulation and the diminution of solvent quantity, high solid coatings have high surface tension. Indeed, in majority, the functional groups of coatings are highly polar and the solvent used in high solid coatings are mostly hydrogen-bond acceptor. All these factors lead to the fact that the surface to be applied must be thoroughly cleaned and properly prepared to reach the highest possible surface free energy.

Shorter Pot Life

Here again, the increased content of functional groups induced side effects and reduce the pot life of high solid coatings [81,84]. Storage stability is affected by the rate of reaction at storage temperature which in turn is influenced by the concentration of functional groups. In high solid coatings, functional groups are more important (at least three times as for the same curing system in conventional coating) because there is less solvent and because functionality is increased to get the right chain extension [81].

Flocculation of Pigment

Avoiding flocculation of pigment is crucial in any type of coating, conventional or high solid. It is not a major problem in conventional coatings but it can be critical in high solid coatings because it can induce an increase in the viscosity of the coating before application [81]. This viscosity increase is attributed to immobilization of continuous phase within and around the irregular surface of the aggregate and to the crowding or packing aggregates with each other [81].

One of the primary factors controlling the stabilization of pigment dispersion is the thickness of adsorbed layer on the pigment particles surface. The low molecular weight molecules in resin for high solid coatings are incapable of providing an adequate adsorbed layer [81].

Cure Window

The cure window for high solid coatings is narrower than for conventional coatings, it is thus important to have a precise formulation and to apply the coating at the right baking temperature and time [81,84,85]. For conventional coatings, the baking temperature, backing time or catalyst amount can be off by ± 10% [81]. There are a large number of hydroxyl groups in each resin molecule and if 10% are not reacted, the change in the properties will be small. However, in high solid coatings, the content of functional groups is optimized to obtain the right final mechanical properties without having a negative impact on the viscosity. Leaving even a small amount of unreacted functional group can, thus, lead to high change in the final properties of the coating [81,84].

Sagging

Sagging is a greater problem with high solids than conventional coatings [81,83,84]. This results from a lower loss of solvent during application for high solid coatings in comparison to conventional coatings. Indeed, a high solid coating having 70–80% solids at the spray gun will have 75–80% solids on the substrate whereas it was 20–30% solids at the gun and 75–95% solids on the substrate for conventional coatings [81,83,84]. The reasons for this difference have not been completely elucidated but result in sagging.

Oven sagging can also happen. High solid coatings tend to sag during the early stages of heating during thermal cure due to the strong temperature dependence of viscosity. It occurs when the increase of viscosity resulting from the crosslinking reactions is insufficient to counteract the decrease in viscosity of high solid coatings as temperature is increased during baking [81,83,84].

It can be seen that, even if high-solid coatings present substantial advantages, the control of the film formation to obtain a defect free film appears more difficult than for conventional coatings. This will obviously have an impact on the performance of these type of coatings.

3.3. *Powder Coating*

3.3.1. Generalities

Powder coatings were first available in Europe in the early 1950s and then in United States in 1972 [87]. The first powder coating was based on polyethylene chemistry and applied by a fluidized bed technology on preheated steel. It is only in the 1960s that powder coatings based on polyesters and epoxies technology were developed. Knudsen and Forsgren [87] stated that powder coatings constituted 11% of the world coating market in 2010s.

In powder coatings, all the constituents of the coating are present in a dry, solventless powder. They are formulated from solid resins, cross-linkers, pigments with small amounts of additives such as UV screeners, flow promoters, catalysts and pigment dispersants. Powder coatings are usually produced by melt mixing the ingredients at elevated temperature under high shear to disperse pigments in a matrix of the major binder components. Even if this process can be either in batches or continuous, the same process steps are found. An example of production line is shown on Figure 11.

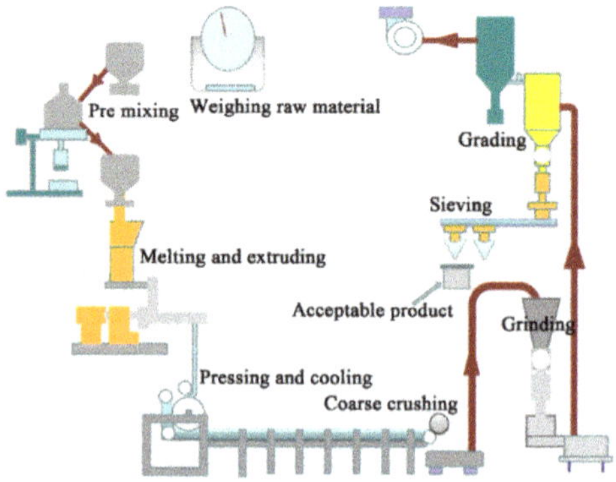

Figure 11. Schematic representation of a production line for powder coating [88].

In addition to their obvious environmental benefits, powder coatings have a number of technical and economic advantages:

- In the case of electrostatic spraying, since the powder is drawn to the substrate by electrostatic charge, a very high ratio of the powder ends up on the substrate. This ratio is higher than the one observed for conventional coating. Furthermore, the little fraction which does not hit the substrate can be recycled in the application booth and sprayed again.
- The high temperature of application (160 to 210 °C) makes it faster to cure and allows the formation of a very dense and highly protective film.
- They are ready to use since no dilution or thinning is needed.
- They are easy to apply because they do not run, drip or sag, unlike conventional liquid coating
- They can cover irregular shapes.

Powder coatings present numerous advantages but the size of the object to be coated is one of the main limitations. Even if the application line allows the coating of quite large objects, this technology is limited to objects and parts [87,89,90]. In addition, whatever the application methods selected, it is harder to change the color of powder coating applied during the application process in comparison to conventional solvent-borne coatings [87,89,90].

Powder coatings can be divided into two main categories as solvent-borne coatings: thermoplastic powder and thermoset powder. The market share for thermosetting coatings is larger (90%) than for thermoplastic coatings [87,89–91].

3.3.2. Application Methods

The method of applications for powder coatings differ greatly from the application methods utilized for conventional solvent-borne coatings and high-solid coatings. That is why it is important to mention them.

Two principal methods are used, namely, electrostatic spray and fluidized bed. Other minor application methods can also be quoted, such as electrostatic fluidized bed and flame spraying. When thin film powder coatings are desired, the preferred application method is by electrostatic spray. For protective and thick-film powder coatings, fluidized bed method is adopted. Before coating, as for conventional solvent-borne paints, the substrate must be carefully cleaned by appropriate surface preparation. Conversion coatings can also be added to improve adhesion and corrosion resistance.

Electrostatic Spraying

Electrostatic spraying is the major process for applying powder even though it was not the first one to be developed. The powder is fluidized in a hopper and conveyed by a stream of air to an electrostatic gun [87,89,90,92]. The substrate is electrically connected to ground so that the charged powder particles are drawn to the substrate by an electrostatic force. The airflow out of the gun has the role to add a velocity toward the substrate to the particles. There are two types of electrostatic gun, the corona and tribo gun, see Figure 12. They differ by their means to input electrostatic charges to the powder.

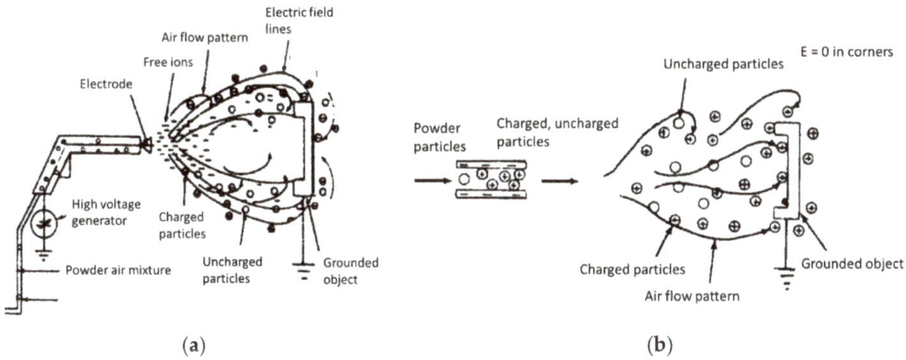

Figure 12. Schematic representation of (**a**) corona charging and (**b**) tribo charging. Reproduced with permission from [93]. Copyright 2002 John Wiley and Sons Society.

Charging the powder by corona effect is the most common method used. It consists of a tube carrying the airborne powder to an orifice at which an electrode is located. The electrode is connected to a high voltage (30–100 kV) and low amperage supply [87,89,90]. Even though the voltage in the gun is high, the current is very low which allow the energy transmitted to the powder to remain quite low. Electrons emitted by the electrode react with molecules in the air, generated a cloud of ions (mostly OH^-), called a corona, around the orifice [87,89,90]. Powder particles come out the orifice negatively charged due to the pick-up of anions and the difference in potential between the substrate and the particles draw them to the substrate surface. One disadvantage of the electrostatic powder spray process using corona discharge guns is that a high voltage field is set up between the gun and the substrate. As field lines will tend to concentrate on the points closest to the spray gun, parts with deep angles or recessed areas will be difficult to coat. This is called the Faraday cage effect.

Another means of applying electrostatic powder employs triboelectric charging of the particles. The particles are charged by the friction generated by streaming through a poly(tetrafluoroethylene) tube in the gun [87,89,90]. Since there is not a large differential in charge between the gun and the grounded substrate, no significant magnetic field lines will be established leading to a minimal faraday cage effect. This will help in the coating of irregularly shaped objects and smoother coating can be obtained with this method. It should be noted that as throughput is slower, stray air currents can deflect more easily the particles between the gun and the grounded substrate [90].

After the application of the powder coating, the substrate is conveyed to a baking oven where the powder particles are fused to a continuous film, flow and crosslink (as a function of the coating type). The volume fraction of sprayed powder that ends up on the substrate is called transfer efficiency which will depend on the geometry of the object and the particle size. In general, the particles that do not adhere to the substrate are particles which received little charges [81]. The particles that do not adhere to the substrate (overspray) are recovered as a dry powder. This powder is usually recycled by blending with virgin powder but this recycling has a limit as the recycled powder will have an increased concentration of particles with poor charging.

The film thickness deposited by electrostatic spraying using the corona gun can be adapted by changing the gun voltage and the distance between the gun and the substrate [94]. To increase the film thickness, the voltage must be increased whereas the gun–substrate distance must be decreased. Thicker film can also be applied by heating the substrate before deposition.

Fluidized Bed

Fluidized bed is the oldest method of applying powder coatings. It is very different from electrostatic spraying since the substrate is dipped into the fluidized bed after being preheated. Once the preheated substrate is in contact with the powder, the powder melts and forms a film. The fluidized bed typically contains a powder container and air supply which are separated by a porous bottom. The powder is transformed into a fluidized state by the passage of clean air through the porous bottom resembling a boiling liquid, see Figure 13. As the thickness of fused particles builds up, the coating becomes a thermal insulating layer so that the temperature at the surface of the coating decreases, finally reaching the stage which further particles do not stick to the surface. The last particles to attach to the coated surface are not completely fused, so the substrate must be moved to another oven to complete the fusion [87,89,90].

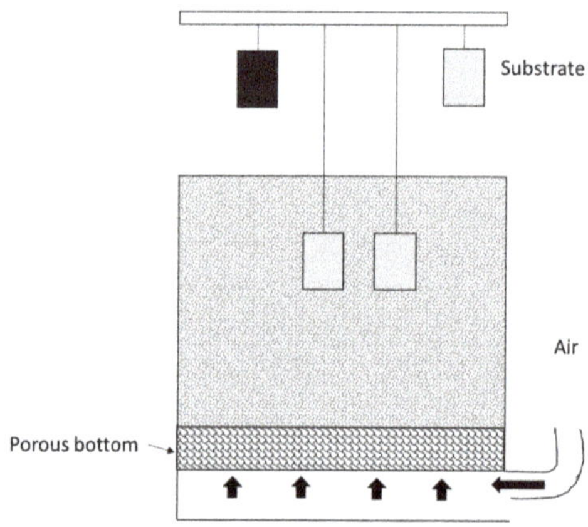

Figure 13. Schematic representation of the fluidized bed process.

The powder applied by the fluidized bed has a particle size distribution ranging from 30 to 250 microns which is larger than particles applied by electrostatic spraying [87,89,90]. Depending on target coating thickness and properties of the powder, the substrate is heated to 230 to 450 °C and dipped for 2 to 10 s. Coatings on the order of 200 to 500 µm are typically produced. The size of the substrate that can be coated is limited because the fluidized bed container must be large enough to readily accommodate them. Most commonly, this method is used for applying thermoplastic coating materials.

Other Minor Methods

Among the other methods which can be quoted, electrostatic fluidized bed and flame spraying can be mentioned.

Electrostatic fluidized bed has a similar process as fluidized bed but electrodes are added to generate ions the air before the substrate is dipped in the powder. As for electrostatic spraying, powder is attracted to the substrate through electrostatic force. This method is used to apply thermoplastic and some thermosetting powder coatings.

Flame spraying is another technology for applying thermoplastic powder coatings. During this process, the powder is propelled through a flame and stays long enough in it to melt and is deposited on the substrate and cooled down before any thermal degradation of the polymer. Only thermoplastic powder can be produced this way due to a short time at high temperature which prevent the completion of a curing reaction [89,90]. In contrast to other techniques, flame spraying can be use on field.

3.3.3. Formulation and Film Formation of Powder Coatings

As for high-solid coatings, the formulation and film formation of powder coatings are more difficult than for convention solvent-borne coatings and a detailed description will be made in this paragraph as it can influence coating performance.

Formulation of Powder Coatings

The formulation of powder coatings is similar to conventional coatings in many regards. The same type of binder, crosslinker and additives can be used. However, their respective ratio in the powder must be optimized to avoid difficulties which could lead to defects in the final paint film [87,89,90]. Another difference between powder and conventional coatings relies in additional additives adapted to the powder production and the application methods. Whereas many additives such as UV absorbers, light and heat stabilizers and slip agents have their basics in conventional coatings, others such as electrostatic additives or post-blend additives are unique to powder coatings [89,95]. For example, degassing additives can be incorporated in the formulation to mitigate the defects caused by volatiles (air, moisture, low molecular weight polymer fragments) emitted during film formation to avoid pinhole defects. Additives to avoid flooding, floating and pigment agglomeration which are used in conventional coatings are not necessary and seldom used in powder coatings.

A significant formulating advantage for conventional coatings is that there are few constraints on the Tg of the binders whereas it can be a crucial parameter for high solid coatings, see Section 3.2.2 An important formulating variable is the ratio of pigments and fillers to the binder in powder coatings [89,90]. It is generally much lower than for conventional coatings since there are no other liquid ingredients present to wet out the pigments. Thus, the pigment volume concentration in powder coatings seldom exceeds about 25% [89,90]. In addition, higher levels of pigments or fillers can reduce the flow out of powder coatings resulting in a rough grainy aspect. Many of the pigments used in conventional coatings are also used in powder coatings. These include inorganic pigments (TiO_2, iron oxides, complex inorganic pigments) and organic pigments. Here again, their concentration in the formulation must be adapted to avoid any negative effects on the powder flow.

For thermosetting powder coatings, the formulation is much harder than for thermoplastic powder coatings. Indeed, the formulation must satisfy a combination of conflicting needs such as the minimization of premature cross-linking during production, stability against sintering during storage, coalescence, degassing and levelling at the lowest baking temperature, cross-linking at the lowest possible temperature in the shortest time possible and film properties that meet or exceed the requirements of the intended application [87,90,96]. In addition, flow and leveling must be balanced to achieve acceptable appearance and protective properties over the range of expected film thickness. For example, coatings that flow readily before cross-linking can form smooth films but they may flow away from edges and corners.

Film Formation

As previously said, after application, the substrate covered by powder is taken to the curing oven. For both classes of powder coatings, the mechanism is first the melting of the powder and formation of a continuous film, Figure 14.

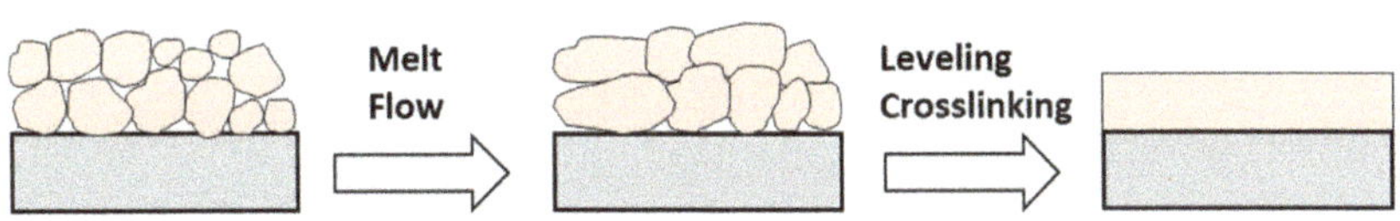

Figure 14. Schematic representation of powder coating film formation.

Thermoplastic and thermosetting powder coatings differ by the fact that a curing reaction happens following the powder melting for thermosetting powder coatings whereas for thermoplastic coatings, the film is formed by the evaporation of air between particles [87,89,90]. This curing reaction complicates the film formation and selecting correct curing conditions is extremely crucial [97].

Two processes can be used to cure thermosetting powder coatings either heat or UV light. Heating is the most common way to complete the curing reaction while UV curing can only be used for specifically designed powder coatings [87]. During heat curing, as the temperature increases the powder starts to melt and float together. The outer powders melt first since the substrate has higher heat capacity and takes longer time to be heated. Curing accelerates as the powder temperature increases and consequently the powder viscosity increases. In heat-curing powder coating, there can be a risk of pinhole formation due to the degassing of air or water molecules (produced by the curing reaction) at the last stage of the curing process when the viscosity is high.

For UV-cured powder coatings, the risk of pinhole defects is lower because the curing reaction happens in two steps. Indeed, powder coatings are first heated until the powder particles melt and coalesce into a molten film. Then, the film is cured by irradiation with UV light in few seconds.

Parameters Influencing Film Formation

There are a lot of parameters which can influence film formation but they can be sorted in two main categories, the ones inherent to the powder particles characteristics (size, distribution and shape) and the ones inherent to the physicochemical and rheological properties of the binder powder (Tg, melt viscosity and surface tension).

The powder particles characteristics will have an impact on the application process, mainly for electrostatic spraying, and on the thickness of the final film. Small particles can induce a problem in the powder transport in the application equipment (electrostatic gun) because they can fill the voids between large particles and lead to powder packing and difficulty in the fluidization of the powder [87,89,90]. These phenomena can result in film thickness variation. In addition, during electrostatic spraying, smaller particles can affect the flow pattern due to the fact that they receive more charge per weight than larger particles [87,89,90]. If too small particles need to be avoided, it is the same for too large particles. Indeed, they can affect the film smoothness. Generally, the largest particles must not be larger than 2/3 of the desired film thickness [87,89,90,98]. In addition, particles shape and density can influence the spray process and powder recovery. Particle size distribution can also play a role on the packing density of the film, the fusion of the particles and the amount of air between particles before fusion. Irregular shapes promote void in the dry powder film.

Concerning the physicochemical and rheological properties of the powder, the melt viscosity and the surface tension of the melt are very important. Viscosity will have an impact on the powder flow and on the coalescence of the powder particles. Generally, viscosity is high immediately after fusion, drops with increasing temperature and finally increases when the coating approaches gelation [87,89,90]. Obtaining the right melt viscosity is essential because too low viscosity will result in sagging, dripping of paint or the blocking of powder particles when stored whereas too high viscosity will induce a poor flow [87,89,90]. The flow, in particular the flow window, will depend on the lowest viscosity attained and by the length of time that the melt powder stays within the right viscosity range. Viscosity is also crucial for the leveling of the film [87,89,90].

Regarding the surface tension, again, the right melt surface tension needs to be found. High surface tension promotes coalescence and the flow of molten powder particles while low surface tension is necessary to wet the substrate [87,89,90]. As for the viscosity, if the surface tension is too high or too low it will generate film defects such as cratering or orange-peel effect (poor flow) respectively. These two parameters will be also linked to the Tg (mostly the viscosity) of the melt which will in turn depend on the binder nature, its molecular weight, its functionality, the curing agent, etc. The Tg of the powder will also be essential for the storage of the powder before application and for the coalescence and leveling of the film. The Tg must be high enough to avoid powder sintering but no too high because it will impact the temperature at which the coalescence and the leveling of the film will happen [81–83]. It is generally preferable that those (coalescence + leveling) happen at the lowest possible temperature.

4. Main Binder Types When Anticorrosive Properties Are Considered

In the previous section, the technology of high-solid coatings and powder coatings was described to highlight their differences in comparison to conventional solvent-borne coatings and to present the challenge to overcome in their formulation. This section is dedicated to the binder typically used in harsh conditions. It includes binder for C5 and Im2 environment according to ISO 12944-5 [99]. First, a general presentation of the binder family will be made and then the particular case of high-solid coatings and powder coatings will be address.

4.1. Epoxy Technology

4.1.1. Generalities

Epoxy technologies are widely used for anticorrosive coating especially because of their excellent adhesion to metal and their high resistance to heat, water and chemicals [2,100–102]. In addition to these properties, epoxy technology exhibits very strong mechanical properties, better alkali resistance than most of other types of polymers but shows a high susceptibility to UV degradation.

Epoxy resin refers broadly to thermosetting polymer that contains an epoxide ring also called an oxirane ring. The most common epoxide used are: diglycidyl ethers of bisphenol A (DGEBA), diglycidyl ethers of bisphenol F (DGEBF), epoxy phenol or cresol novalac multifunctional resins for very aggressive environments, see Figure 15.

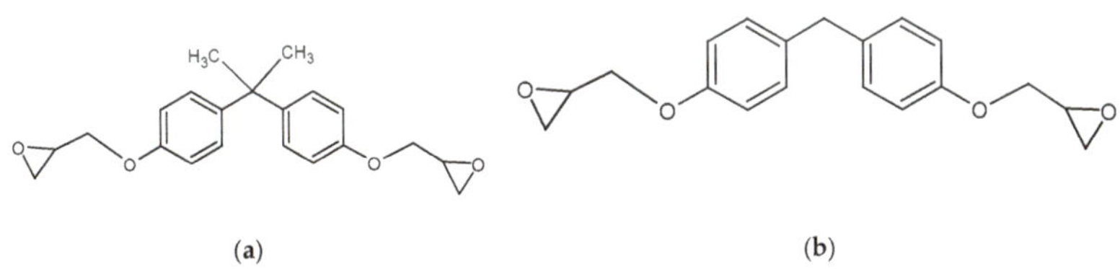

Figure 15. Basic chemical structure of (**a**) diglycidyl ethers of bisphenol A (DGEBA) and (**b**) diglycidyl ethers of bisphenol F (DGEBF).

It is accepted that the corrosion protection available with epoxy technology comes from a combination of factors, i.e., the presence of aromatic group in the bisphenol structure which form a good barrier protection and the presence of secondary hydroxyl group along the chain for adhesion purpose [101]. However, the presence of aromatic group in the bisphenol structure is also what makes epoxy coating highly susceptible to UV degradation. It is because of this susceptibility that chalking and yellowing are often encountered for epoxy coating, see Figure 16. Epoxy coatings are often used as primer or intermediate coats and combine with another type of chemistry for the topcoat.

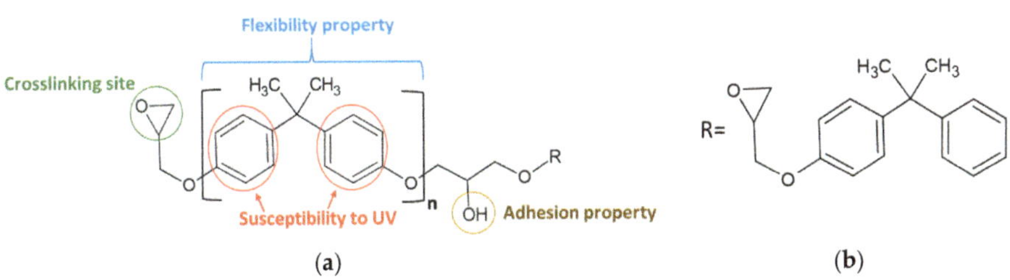

Figure 16. Various properties of epoxy resin as a function of the chemical groups considered (**a**) Detailed chemical formula and (**b**) Detailed chemical formula for the functional group named R.

The crosslinking of epoxy resin usually occurs by the opening of the ring by curing agents (generally amines) through a proton donor reaction. The high strain due to bond angle encountered in the oxirane ring makes it highly reactive. There are a number of curing agents available to react with epoxide; amines (aromatic or aliphatic), mercaptans, amides, phenols, isocyanates, anhydrides acids, etc. Thus, the structure of the epoxy coating will partly depend on the nature of the curing agent, see Figure 17. The oxirane ring being highly reactive, most of these curing agents can react rapidly and at room temperature as soon as they are mixed together.

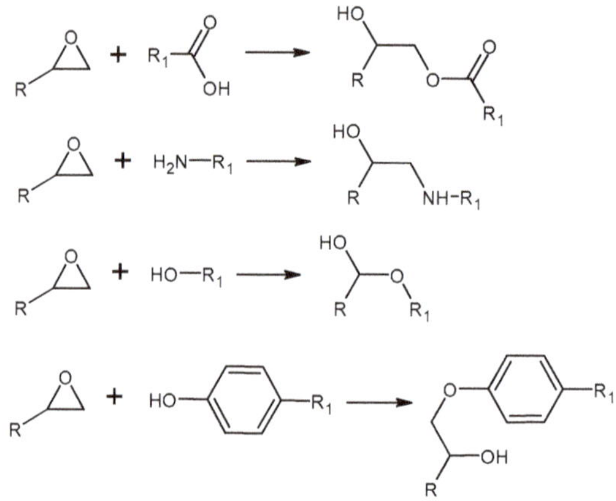

Figure 17. Typical reactions of the epoxide group to form epoxy coating as a function of the curing agent.

It is the molecular weight of the resin which will largely dictate the coating properties, see Figure 18. For example, increasing the molecular weight will increase the flexibility, the adhesion or the toughness with the counterpart of decreasing the crosslink density or the solvent resistance [1]. The difference in final properties as a function of the molecular weight comes from a difference on the number of functional groups per unit, i.e., low molecular weight possesses more functional groups. Resins of differing molecular weight are usually blended to obtained balance property as a function of the application needed.

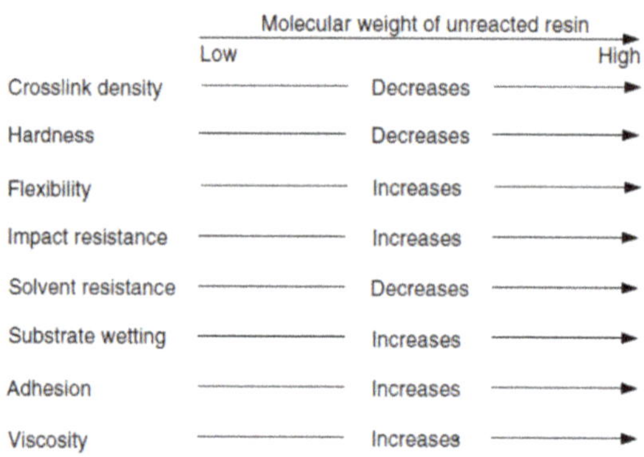

Figure 18. General characteristics versus molecular weight of epoxy coating. Reproduced with permission from [1]. Copyright 2009 Springer Nature.

It should be noted that the choice of the curing agent will also impact the final property of the coating. For instance, epoxy coatings formed by reaction with polyamides are more flexible than epoxy coatings formed with amines. Polyamines are relatively small molecules which will lead to tight crosslinking whereas polyamides will allow crosslinking to occur lengthwise due to their much larger size.

4.1.2. High Solid Epoxy Coatings

The use of reduced molecular weight oligomer leads to the use of standard bisphenol A epoxy resin modified with dimer acids. In theory, two molecules of diepoxide resin can be joined with one molecule of dimer acid [84]. Thermosetting acrylic resin oligomer can also be used to modify epoxy high solid coatings. The products are generally formulated and applied at much higher solid levels [84].

Concerning amine cured epoxy coatings, to be used as high solid coatings, it is common to increase both the extender level and the thixotrope additive level to avoid sagging of the coating during application [83,84].

The use of low molecular weight in epoxy high solid coatings can lead to more brittle coatings than conventional coatings [1,84,103]. Thus, they can be more susceptible to damage from impact, abrasion gouging. All these types of damage can compromise the barrier properties of the coating. One way to remedy this problem is by the addition of flexibilizers or toughening agents [103].

When considering high solid primer, typically epoxy type for their good adhesion, they are more difficult to formulate due to possible pigment flocculation as depicted in the previous section.

4.1.3. Epoxy Powder Coatings

They are the oldest and still one of the largest classes of thermosetting powder coatings. One explanation is that during the development of thermosetting powder coatings, solid epoxy resins and well-known latent curing agents were already available leading to a quicker development. The most common crosslinker used in powder curing is dicyandiamide (DICY). This crosslinker leads to complex polymeric structure due the complexity of the polymerization reaction which is still not completely elucidated [104–107]. DICY can be modified to enhance its solubility in epoxy and improve the film uniformity. As the curing reaction between DICY and epoxy can be slow it is common to add a catalyst such as imidazole molecules [83,89,108]. Other typical curing agents are phenols, anhydride, aromatic amines and aliphatic diamines.

The same properties encountered for conventional coatings are found for epoxy powder. Epoxy powder coatings are mostly used for corrosion protection in particular for onshore and offshore oil and gas pipelines which are their largest market. In the pipeline coating industry, this type of coating is called fusion-bonded epoxy (FBE) [87,108,109]. It can be used as a single coat and as the first layer in a multi-coats system [110]. Typical binder compositions include a solid epoxy resin and a dihydrazide crosslinker.

4.2. Polyurethane Technology

4.2.1. Generalities

Unlike epoxy coating, polyurethane coatings are known to be highly UV resistant and made them the favorite choice for topcoat paint but they can also be available as primers or intermediate coats. Polyurethane coatings exhibit excellent water resistance, good resistance to acids and solvents, better alkali resistance to most polymers and in general good mechanical properties in addition to their good weathering properties [1,100,111,112].

They are formed by isocyanate (R–N=C=O) reactions with active or labile hydrogen atoms coming from, for example, alcohols, amines or water molecules, see Figure 19. The crosslinking of polyurethane coatings will also depend on the type and configuration of both the isocyanate and curing agents, temperature and presence or not of catalyst.

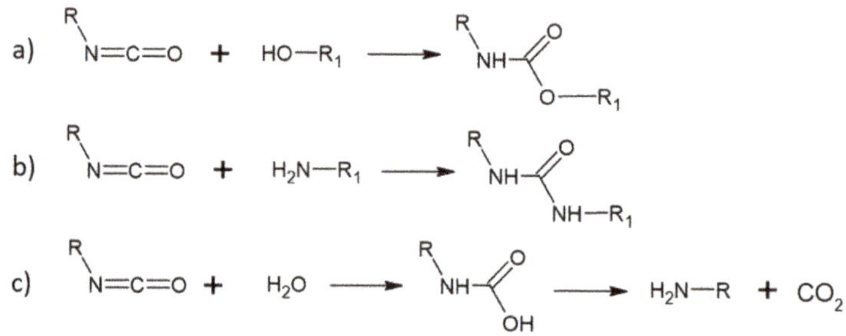

Figure 19. Some typical isocyanate reactions (**a**) hydroxyl reaction, (**b**) amino reaction and (**c**) moisture core reaction.

Two types of polyurethanes can be distinguished as a function of their curing mechanisms; moisture-cure urethanes or chemical-cure urethanes. Moisture-cure urethanes are single pack coating and chemical-cure urethanes two components pack coatings. In single pack coating, the isocyanate group reacts with any labile hydrogen, moisture in the air included. The use of moisture-cure urethanes is limited to by the concentration of moisture at the surface which can induce adhesion problems and bubble formation due to the presence of carbon dioxide as by-product of the cure reaction.

Both single pack or two components pack coatings can be made of either aliphatic or aromatic isocyanates as a function of the final property needed. Aromatic polyurethanes, though less expensive, have more chemical and solvent resistance but are more susceptible to UV degradation due to the aromatic structure, such as with epoxy technology [1,100,111]. It is what makes them preferably used as primers or intermediate coats. Aliphatic polyurethanes can be linear or cyclic and offer better UV resistance and are used as topcoat when good weatherability is needed but are more expensive and react more slowly [1,100,113,114]. Figure 20 shows some typical chemical structure of aromatic and aliphatic isocyanates.

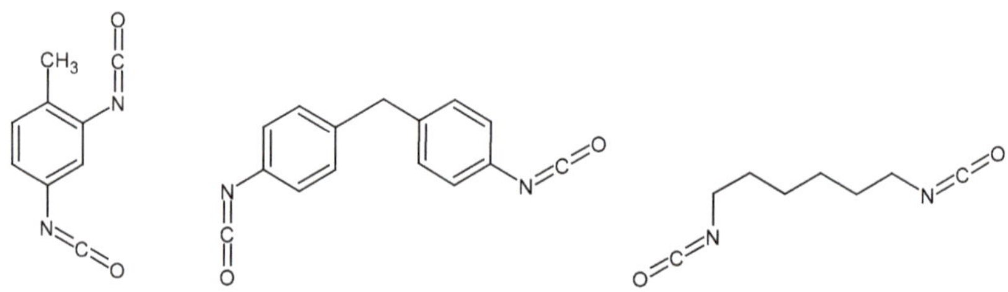

Toluene diisocyanate (TDI)	Methylene diphenyl 4,4′-diisocyanate (MDI)	Hexamethylene diisocyanate (HDI)
(**a**)	(**b**)	(**c**)

Figure 20. Some typical isocyanate molecules (**a**) Toluene diisocyanate, (**b**) Methylene diphenyl 4,4′-diisocyanate and (**c**) Hexamethylene diisocyanate.

Aliphatic polyurethanes are mostly polyester and/or acrylic polyol whereas aromatic polyurethanes are mostly polyether polyol.

4.2.2. High Solid Polyurethane Coatings

The low viscosity, high functionality and high isocyanate content of the oligomeric polyisocyanates make polyurethane especially suitable for high solid coatings. There is an ASTM standard for describing the different types of polyurethane and the characteristics of high solid and solventless polyurethane coatings are detailed in the ASTM D16-type V [115]. In the development of high solid polyurethane coatings, as for epoxy coatings, the adopted approach is to lower the binder of the resin, add diluents or lower the viscosity of the polyisocyanate crosslinking agent [84,111,116,117]. Renz et al. [118] described the effect of these modifications and their possible negative effects. Hydroxy-terminated polyester and hydroxy-functional acrylic resins are the most common polyols used in the formulation of high solid polyurethane coatings [84,116]. Other way using self-crosslinking isocyanate system can also be found [116]. Depending on application area and mode of application, the solid level for typical two package polyurethane coatings range between 55–70% weight solids and 40–60% volume solids [116]. Still in the spirit of achieving greener chemistry, high solid polyurethane coatings can be formulated with vegetable-based polyols [119].

4.3. Powder Polyester Technology

They are mostly used for lining applications due to their ability to be applied as very thick film builds. They have good solvent and chemical resistance but no resistance to alkaline conditions because of the sensibility of the ester linkage. As they exhibit also excellent water resistance, they are often used for seawater applications. The principal class of polyester powder coatings is based on acid functional, saturated polyester resins cured using triglycidyl isocyanurate (TGIC) [87,89,90,120]. Even though TGIC has been the dominant curing agent for polyester resins, it is being slowly replaced by primid due to its hazardous nature for environment and health issues. In case of TGIC cured resins the binder/crosslinker ratio is 93:7 due to a high equivalent weight and a low degree of functionality of the binder. Polyester powder coatings possess good adhesion, protective properties and are less susceptible to UV degradation due to the absence of aromatic molecule in their formulation. However, they can be sensible to the accumulation of moisture at the surface which can lead to hydrolysis, oxidation and degradation of the resin [87,89,90]. Typical products coated with polyesters include architectural components, agricultural equipment, household appliances and electrical enclosures.

4.4. High Solid Polysiloxane

Polysiloxane is any one of a class of synthetic materials characterized by silicon and oxygen linked together in the siloxane bond, having organic group attached to silicon which have a distinct impact on the final properties of the coating [100,121,122]. Siloxane can be chemically reacted with epoxy, acrylic or other organic groups which can lead to the formulation of a broad variety of coating types [100,121,123]. Two main factors have made polysiloxanes coatings highly valuable for use in protective coatings; their strong chemical bond in comparison to typical carbon bond of organic binders and their low surface energy. Their strong chemical bond leads to high heat resistance and long-term weather resistance [100,121]. Their low surface energy has an impact on their hydrophobicity properties, wetting properties and can improve flow and leveling during film formation [121].

Due to the fact that silicon can be from mono to tetra functional, see Figure 21, and due to the chemical nature of the functional group, it can lead to a large panel of properties. For example, more trifunctional silicon results into coating with an improved hardness but with brittle behavior and with an improved compatibility with organic polymers.

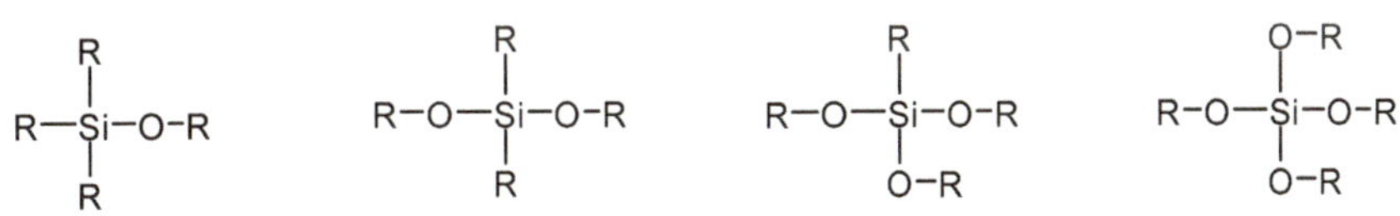

Monofunctional Di-functional Tri-functional Tetrafunctional

With R = hydrogen, methyl, ethyl, propyl, octyl, phenyl or other organic groups (epoxy, acrylic, urethane, etc.)

Figure 21. Example of multifunctional siloxane structure.

As for conventional coatings, changing the crosslinking density of polysiloxanes will result in different properties; high crosslinking will lead to strong, hard, protective and chemically resistant coating but quite brittle while low crosslinking density will result in more flexible coating but with worse barrier properties [100,121]. The latter will be more suitable as a topcoat to be applied on a protective epoxy, for example, while the former will be a UV-resistant barrier coating. Due to limited surface tolerance, this coating cannot be used without a primer. Polysiloxane coatings are an alternative choice to polyurethanes coating. Previously reported accelerated weathering data has fingerprinted polysiloxanes technology as being a coating with extremely good weatherability. Florida exterior weathering data (45° South, five-years continuous exposure) validated that polysiloxanes coatings are very durable. This data is proof that siloxane coatings have excellent aesthetic weathering attributes in terms of gloss retention and chalk resistance properties [124].

Polysiloxanes can be formulated as both single or two component(s). As similar for polyurethane coatings, the curing of one component coating requires the presence of humidity in the atmosphere whereas in two components coatings, the silicon chain is formed in the resin and the curing is achieved by a silane with desired functional groups. Inorganic ratios can vary from 37% to 77% (weight by weigh) but is typically around 60%.

4.5. Other Technologies

In addition to epoxy and polyurethane binders, several other chemistries can be found in the area of anticorrosive coatings. Among them acrylic and alkyds coatings can be quoted. For more detail information about these coating technologies, see [4,100].

4.5.1. Inorganic Zinc Silicate Coating

Zinc level of 90% or higher are common in inorganic zinc silicate coating which makes it almost entirely composed of zinc pigment providing cathodic protection. The amount of binder is optimized to keep the electrical contact between the zinc particles and the surface. As the name of the coating implies it, the binder is an inorganic silicate which may be either a solvent-based, partly hydrolyzed alkyl silicate or a water-based, highly alkali silicate. The general characteristics of these coatings are the ability to tolerate higher service temperature, excellent corrosion properties, the requirement of top-coating for specific pH environment and very thorough surface preparation [125].

4.5.2. Other Types of Powder Coatings

Thermosetting Powder Coatings

Polyester-epoxy hybrid powder coatings combine the decorative properties of the polyester and the toughness of the epoxy. They were developed to be an alternative to anhydride cured epoxy powder coatings. Despite the presence of polyester, these hybrid coatings are still susceptible to UV degradation and are preferred to be used indoor [87,89,90]. They are typically used for furniture and kitchen appliances.

The application segment for polyurethane powder coatings is low. Generally, blocked isocyanate group approach is used to prevent premature crosslinking.

Thermoplastic Coatings

The two most important vinyl coatings are polyvinyl chloride (PVC) and polyvinylidene fluoride (PVDF). Vinyl polymers belong to a group of resins having a vinyl radical as the basic structural unit.

Polymerization of vinyl chloride into PVC homopolymer or its co-polymerization with different co-monomers is carried out by a free radical mechanism. Due to its brittle nature, PVC powder coatings are plasticized formulations. The use of plasticizers also improves their impact resistance but at the same time reduces their tensile strength, modulus and hardness [87,89,90]. Thus, the choice of the amount of plasticizer is crucial to obtain the right balance between hardness and flexibility. PVC is degraded by UV light and therefore is preferred for indoor use. They are mechanically tough and protective when applied in sufficient thickness. Their partial crystallinity contributes to the stabilization of the powder against sintering.

Fluoropolymers, such as PVDF are used for coatings requiring exceptional exterior durability and enhanced resistance to corrosive environment. Most of them are modified with thermoplastic acrylic polymers or other fluoropolymer resins containing comonomers to improve melt flow and application characteristics [87,89,90]. In addition to their enhanced UV resistance and gloss retention, they exhibit good abrasion resistance. Their low surface energy also makes them dirt repellent. They are generally applied as topcoats mostly for decorative purpose in architectural applications.

Polyolefin powder coatings consist mostly in polyethylene and polypropylene chemistry. They give rise to coating with a smooth finish with an almost waxy feel. Their chemical composition i.e., no side chain functional groups allow them to have almost no water absorption and a very strong resistance to chemicals and solvents [87,89,90]. However, this inert characteristic is a drawback when adhesion to the substrate is considered. Indeed, there is no functional groups to establish primary or secondary bond with the surface substrate. Thus, they usually need the use of a primer or the incorporation of other monomers in their formulation, such as acrylic acid. This type of powder coating is frequently used on lab equipment that require regular washing.

Polyamides powder coatings relies mainly on Nylon 11- and Nylon 2-chemistries and the properties of these two types of nylon are similar. Polyamide powders can be prepared either by melt-mixed or dry-blend process as a function of the final gloss wanted (higher gloss for melt-mixed process) [87,89,90]. These coatings are mechanically tough, tolerate high temperature and resist to a wide range of chemicals and solvents. They have also a pleasing appearance and are suitable for food contact. They are mostly used for the coating of handles, wire goods, automotive parts and medical equipment.

5. High-Solid and Powder Coatings for Marine and Offshore Application

It was important to define the type of binder which can be utilized for designing anticorrosive coatings because it is the binder which will have an impact on the overall coating properties for the most part. Before reviewing the behavior of high-solid coatings and powder coatings in marine and offshore application, it is essential to define the type of stress which can be encountered. This part is directly linked to degradation mechanisms describe in Section 2.4.

5.1. *How to Define Marine and Offshore Applications?—Stresses Encountered*

The definition of the corrosive environment to which anticorrosive coatings can be exposed is reported in the ISO 12944-2 standard [126]. It divides the environment into three exposure types: immersion, atmospheric and splash zone, the immersion and atmospheric exposure having subclasses. Each type of corrosive environment has its own set of stress factors which will lead to different types of degradation mechanisms. Thus, the specific requirements for anticorrosive coatings are naturally highly dependent on the environment and elements that the coating may experience during service and the time of exposure. Regarding marine and offshore applications, marine atmospheric exposure, immersion in seawater and splash zone need to be considered.

5.1.1. Atmospheric Exposure

Atmospheric corrosion is usually defined by being the result of the interaction between a material and the surrounding atmosphere. Even though, this definition seems rather simple, it implies a relative complex problem. Indeed, the composition of the atmosphere is far from being simple and it evolves as a function of physical parameters such as the temperature, the humidity, presence of gaseous atmospheric contaminants and the presence of aerosols.

Concerning aerosols, they come from sea-spray aerosols produced from waves through bubble bursting by breaking waves at the ocean or shore. The residence time of aerosol particles varies from a few minutes to days as a function of the frequency of the rain and the orientation of the materials (allowing rinsing of the surface). The deposition of chloride from sea spray aerosols depends largely on the distance from the sea shore and wind direction. As explained in Section 2.4, marine aerosols can produce species which have a preponderant role in the corrosion process. Regarding atmospheric gases, mostly sulphur dioxide can be quoted. It is one of the most corrosive agents in industrial and urban atmospheres. Sulphur dioxide has a strong corrosion stimulating effect and it was suggested that the oxidation of SO_2 to form sulphate leads to a pH decrease, a dissolution of surface oxides and an increase corrosion.

To be able to interact with the material, both the gases and the aerosols will need to be dissolved and thus will need the presence of water at the metal surfaces. The presence of water at the surface of the material will have an impact on the solubilization of the contaminants and on the properties of the paint layer [127,128]. Thus, the relative humidity and temperature appear as the most important to consider. In addition, the wet and dry cycles which undergo the materials are of importance as climatic parameters are never under stationary conditions with constant humidity and temperature during outdoor exposure. When considering the day/night pattern, it is obvious that condensation on the material and thus the electrolyte thickness will be largely impacted. The temperature will also have an impact on the rate of chemical and electrochemical reactions and on the diffusion of pollutants through the coating layer, as specified in Section 2.4, and will play a role in coating aging as depicted in Section 2.4. The pollutants solubility will also be impacted by the temperature e.g., O_2 solubility decreases with the increase of temperature [129].

In addition to the climatic parameters and the atmosphere pollutants such as UV radiation, the frequency of rain, the distance to the sea, the wind force or direction will need to be considered.

In conclusions, coatings applied in atmospheric environments will be subjected to locally alternating conditions with respect to ultraviolet degradation, heat, moisture, salt and gas concentration. As a function of stresses encountered, the mechanism of coating failure will be different as depicted in Section 2.4.

5.1.2. Immersion

At the difference of atmospheric corrosion which can only occur when a liquid film is deposited on the surface, in immersion conditions, the corrosion rate is determined mainly by the amount of oxygen that reaches the cathode, whereby the electrolyte resistance is of no consequence due to the large cross-section and good conductivity. Due to good conductivity of the medium, the anode and cathode can be some distance apart in seawater.

In the case of immersed coating systems, the aggressiveness of the environment is very specific. The overall corrosivity of the environment depends on a combination of parameters such as temperature, salinity, pH and content of dissolved gasses (mainly oxygen) and a combination of factors from different origins (chemical, physical or biological), see Table 2.

Table 2. Main seawater corrosion parameters [129].

Chemical Factors	Physical Factors	Biological Factors
Soluble salts	Temperature	Fouling
Solute gases (O_2, CO_2)	Movement	Animals (consumption or production of gases)
pH levels	Pressure	Plants (consumption or production of gases)
-	Solids	-

The salt content of seawater is defined as the total mass of salt solution in 1 kg of seawater. Far from the land in the major oceans, it has an average salt content within the range of 32–37.5 g/kg. The salt content can deviate from this range in particular cases, such as marginal or enclosed seas. Considering all the types of seas, it can be concluded that seawater has a salt content levels of approximatively 35 g/kg and a chloride ion content of 19 g/kg resulting in the seawater composition. The primary cations in seawater are sodium (10.78 g/kg), magnesium (1.29 g/kg), calcium (0.41 g/kg), and potassium (0.39 g/kg) whereas the primary anions are chloride (19.38 g/kg) and sulfate (2.71 g/kg) [129].

The aggressiveness of seawater near coast can be more important due to the presence of gases such as carbon dioxide and sulphur dioxide coming from human activities (emissions from industrial areas). In addition to emission of gases, the presence of industrial areas can raise the amounts of ammonium salts, sulfides and mordant residues and thus increase the aggressiveness of seawater [129]. The salt content will also impact the physical properties of seawater, such as the density, the freezing point, vapor pressure and electrical conductivity. Concerning electrical conductivity, as the salt content and temperature rise, it will drop accordingly.

The solute oxygen content is of primary importance as a component influencing the corrosion behavior in seawater. The content of solute oxygen depends mainly on temperature and salt content. It drops when temperature and salt content increases. The gases dissolved in seawater do not reach saturation levels, not even close to the surface.

The carbon dioxide content of seawater is determined in water close to the surface by exchange with atmospheric carbon dioxide. The solute carbon dioxide influences the balance of carbonate/hydrogen carbonate and thus the formation of lime deposits or covering layers which can have an influence on the corrosion rate.

It is important also to consider fouling phenomenon because it can initiate or exacerbate corrosion processes. It can also have an economic impact when considering fuel consumption. Antifouling coatings represent a particular case of marine coatings and is still a hot topic as regulations about biocides are still evolving [130–133]. Fouling comes from the colonization of material surfaces, constantly or frequently in seawater, by organisms (flora or fauna). The most common components when considering fouling aspects and deposits are bacteria, algae, barnacles, tang, bryozoans, serpulids, tunicates and molluscs.

Others parameters, such as cathodic protection and formation of calcareous deposits on defects should also be considered as they will have an impact on the corrosion rate.

5.1.3. Intermediate Zones

In the previous sections, it was shown that the factors influencing atmospheric corrosion and corrosion due to immersion were different and distinct. In the case of intermediate zones, such as tidal or splash zones, all the factors mentioned above should be considered. Corrosion in intermittently wetted surfaces is significantly more severe than under fully atmospheric and fully immersed conditions [134,135].

Indeed, in the splash zone, the materials are under the influence of high moisture levels and high salt content levels. Stress factors from solar exposure and UV light are also of consequence however no fouling takes place in the splash zone. In tidal zone, unrestricted oxygen access to the moist surfaces in the low water phase enhances corrosion. In this zone, founding and wave movements can also have a considerable influence on corrosion processes.

5.2. Coating Behavior

The different stress factors were described in the previous section. It was shown that they differed as a function of the aimed exposure environment. Thus, an anticorrosive coating designed for marine atmosphere will not necessarily be optimized for immersion exposure. Generally, the application field of high solid coatings is for marine application, such as anticorrosive coating for ships or infrastructure near coast whereas the application field for powder coatings is for offshore application, in particular for the protection of pipelines.

5.2.1. High-Solid Coatings

A typical anticorrosive system for highly corrosive marine environments usually consists of a primer, one or several intermediate coats and a topcoat. The function of the primer is to protect the substrate from corrosion and to ensure good adhesion to the substrate. The function of the intermediate coat is generally to build up the thickness of the coating system and impedes transport of aggressive species to the substrate surface. It should also ensure good adhesion between the primer and the topcoat. The topcoat is exposed to the external environment and must provide the surface with the required color and gloss. The overall performance and durability of a coating system is very difficult to assess because it will depend on the properties of each coats, interaction between coats, adhesion to the substrate, application process and how the global system respond to stress.

Typically, when considering an anticorrosive coating for a harsh marine application and a high durability, according to ISO 12944-5 standard, epoxy and polyurethane chemistries are advocated due to their great anticorrosion properties as described previously [99,136]. Generally, the epoxy technology is used for primer and intermediate coats and polyurethane mainly as the topcoat for the reasons explained previously. Both these types of technology are available as high-solid coatings as depicted in Section 4.1 and Section 4.2. In addition to polyurethane, high-solid polysiloxanes coatings can be used as topcoat as mentioned in Section 4.4.

It is not unusual to have an anticorrosive coating not entirely designed with high-solid coatings but with a combination of traditional solvent-borne coating [137,138]. Typically, a solvent-borne primer coating is used with high solids intermediate coats and topcoats. As the same chemistry as for solvent-borne coating is used in case of high solid coating, the same advantages and drawbacks attributed to their chemistry "class" will be found. Indeed, an epoxy solvent-borne coating and an epoxy high-solid coating will both, for example, exhibit an acute sensibility to UV degradation but improved adhesion with the steel [1,100,101].

As, generally, a combination with of solvent-borne and high-solid coatings are found, it is also rather difficult to firmly say that high-solid coatings are better than solvent-borne coatings. That is why only examples of these types of coating performance will be developed in this section. The only feature of high-solid which could lead to a lower performance with regards to solvent-borne coating is their tricky application which can result in defect in the coating film and thus premature failure, see Section 3.2.2. They cannot be applied as easily as conventional solvent-borne coatings and it can be an explanation in the fact they are still less used traditional solvent-borne coatings. The importance of the application step and its possible difficulty for high-solid coatings is critical when considering the performance of these types of coatings but few information clearly linking them were found. It should also be noted that the presence of low molecular weight/low crosslinking density microregion can be weak points because they can be considered as low ionic resistance areas allowing the transportation of corrosive agents [139]. However, they can offer improve barrier coatings as higher film thickness can be achieved more easily. Indeed, barrier coating properties can be improved by the increase of the thickness of the coating [1].

Diaz et al. [140] studied the performance of two anticorrosive coatings with an epoxy-siloxane topcoat and compared it to the performance of a traditional anticorrosive coating with a polyurethane topcoat. They did several accelerated tests, such as a salt spray, Prohesion and Kersternich test. In addition, they studied the coating permeability, adhesion and electrochemical properties. They found that the painted systems with an epoxy-siloxane topcoat exhibited better anticorrosive properties than traditional coatings. They attributed this behavior to their lower water vapor and oxygen permeability. It was shown that polysiloxanes topcoat could show some cracking due to the freezing phase at −20 °C in the cyclic test recommended in the ISO 12944-6 standard for the qualification of anticorrosive coatings [141]. This cracking could lower the barrier properties of the overall system. Signs of cracking were also observed when comparing the anticorrosive properties of an epoxy siloxane hybrid coating and a traditional two coats epoxy-urethane system [137]. The modification of the epoxy coating with siloxane did improved the weathering ability of the coating without lowering its anticorrosion properties. However, due to the higher crosslink density of epoxy-siloxane hybrid coating, cracking was happening. A remedy such as the addition of elastomer can lower the apparition of cracks [122]. Cases of polysiloxane topcoats peeling were also reported on ship due to problems in flexibility property of the polysiloxanes topcoat [142]. Case of failure with two-coat systems (zinc epoxy primer and polysiloxane topcoat) on an offshore platform was documented [143]. It was attributed to low film thickness, inevitable on these types of structure, rather than to a specific weakness of the polysiloxanes. Good behavior of painted systems with polysiloxane topcoats was also reported in a cyclic accelerated tests and outdoor exposure in different sites [138]. However, this behavior could be attributed to the presence of a zinc rich primer.

Husain and Madaj [144] studied the performance of an epoxy and polyurethane high solid coatings intended for marine industrial atmospheres. They performed a salt spray test, performed infrared and electrochemical analysis. They showed that initially the epoxy coating was very compact and adherent to the steel but a high diffusion of water through the coating was happening after 500 h of exposure. They attributed it to localized micro-corrosion events. Despite this diffusion of water, the epoxy system still exhibited the best performance. Concerning the polyurethane coating, it also showed degradation with the time of exposure. However, it seemed more tolerant to chloride contamination. Shreepathi [145] also concluded about the good performance of a high solid epoxy coating when tested in a neutral salt spray and monitored by EIS. They attributed this good performance to the presence of zinc salt as anticorrosive pigment. These pigments led to an increase of the coating resistance due to the formation of non-conductive corrosion products. Good performance due a high solid polyurethane topcoat in outdoor exposures was also described by Santos et al. [146]. They studied the behavior of several anticorrosive coatings having different combination of coats playing on several types of chemistries and properties. Among the systems studied, a hybrid coating with a waterborne primer, an epoxy high solid intermediate coat and a high solid polyurethane topcoat was selected. In addition to outdoor exposure in a C5 site with high UV radiation, they performed a salt spray test and a condensation test according to ISO 12944-6 standard. They observed that this coating was the second-best coating when considering outdoor exposures with good gloss retention and low chalking mostly due to the polyurethane topcoat. Only one system with traditional solvent-borne coatings exhibited better performance however this coating had a zinc ethyl silicate primer which was responsible for its low degradation. These observations showed the crucial role of the primer when considering the overall performance of an anticorrosive coating. Sanchez-Amaya et al. [147] studied the behavior of epoxy coating, applied on steel with a 300 μm thickness, immerged in HCl 3M media during 14 days. In addition to assessment of the panels after test, they studied the degradation of this coating by electrochemical impedance measurements and electrochemical noise. They noted after the test the apparition of heavy blistering due to the development of anodic blistering through the chemical degradation of the coating. By electrochemical measurements, they showed that, initially, the coating had a good anticorrosive behavior but that after two days of exposure in the media it began to failed and eventually lead to the development of heavy blistering.

One other parameter when considering the use of high solid coating as solution for the protection of steel, in particular, ballast tanks, it their advantages in term of coating's edge retention [148]. Guan et al. [148] showed that high solid polyurethane could be an effective solution in these cases.

In conclusions, zinc-based primers or high solid epoxy primers in combination with polyurethane (high solid or not) or polysiloxane topcoat have been the state of the art in corrosion protection and weathering resistance.

5.2.2. Powder Coatings

Typical Powder Coatings Use in Offshore Application

Coatings used for offshore application are traditionally applied on pipe immerged in seawater or soil. In addition to be resistant to water, they need to resist to cathodic protection applied on pipe. When considering powder coatings used in this application segment, fusion-bonded epoxy (FBE) is the preferred choice. It can be applied as a single layer or as the first layer in a multi coats system [87,110–112,149]. Kehr has written a comprehensive review on the use of FBE for corrosion protection of pipelines [110].

Before reviewing the performance of FBE as anticorrosive solution for pipelines, the procedure of application of FBE on pipelines will be presented because it differs from traditional production lines. The process for coating pipe spools is illustrated on Figure 22.

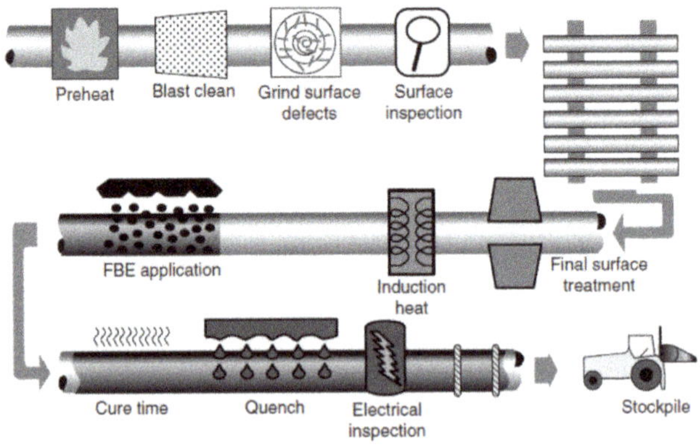

Figure 22. Application of fusion-bonded epoxy (FBE) on oil and gas pipelines. Reproduced with permission from [150]. Copyright 2008 Elsevier.

The pipe is first preheated and blast cleaned. If desired, the surface can also be phosphated to improve the anticorrosive performance. The steel pipe is then further heated by induction heating to a temperature of about 230 °C, before the powder is applied. The typical temperature range for the application of FBE is between 180 and 250 °C [87,149,151]. Since the powder is applied on a preheated surface, the powder immediately melts, flows together into a continuous film and starts to cure. The pipe spool may then receive additional coats of thermally insulated materials, such as polyethylene or polypropylene.

Performance of Powder Coatings

As for what was described in the previous section, the fact that powder coatings are based on the same chemistry as traditional solvent-borne coatings, they present the same advantages and drawbacks inherent to their chemistry "class". Thus, in the case of epoxy type coating, an excellent combination of mechanical properties, chemical stability, corrosion and wear resistance and adhesiveness are observed [112,151–154]. However, they present a high UV sensibility. Even, if pipelines are not necessarily exposed to UV degradation, it can be an important factor to take into account. Indeed, before their use on a pipeline structure, pipes coated with FBE can be stored outdoor and thus suffer from UV degradation which could lower their barrier or other properties before use [155]. Cetiner et al. [155] studied the effect of preliminary UV exposure on FBE in stockpiled pipes on their gloss, cathodic disbondment, adhesion, flexibility and impact properties. The outdoor exposure was carried during 15 to 21 months in different positions as of solar exposure. They showed that UV degradation had a significant impact on gloss and flexibility properties but low impact on cathodic disbondment, adhesion and impact properties.

Several studies have reported that FBE, as a single layer coating, exhibited excellent adhesion to well cleaned steel, good chemical resistance, good flexibility at low temperature (−30 °C), very good cathodic protection properties and good abrasion resistance [153].

These studies also highlighted some drawbacks typical of FBE coating when applied as single coat [156,157]. Indeed, as for all powder coatings, FBE effective adhesion to steel is dominated by the surface preparation and a more careful surface preparation is needed in case of powder coatings, see Section 3.3.3 [84,87,89,151]. Papavinasam [157] showed that defects observed, such as blisters formation, after five years of service for FBE coated pipes mostly came from a non-optimal surface preparation.

Another typical problematic encountered in the case of FBE coated pipe is the adhesion between the coating and the steel pipe [156,157]. Even if, epoxy type coating exhibit excellent adhesion properties it can still be a problem, such as in case of poor surface preparation or due to the easily water transportation to the epoxy coating through the presence of hydrophile groups [158,159]. One way to improve the adhesion between FBE and steel is to apply a surface treatment before coating the pipe with FBE, as similar for all typical anticorrosive purposes [156–158,160–162]. Among the solutions proposed, surface treatment based on zirconium, silane, neodymium-based nanofilm or cerium can be quoted [156,158–162]. Ramezanzadeh et al. [156] showed that treatment of steel substrate by cerium compounds could be an effective approach for enhancing epoxy coating adhesion and corrosion properties. They observed an improvement in the resistance to cathodic delamination due to the covering of cathodic sites on steel with a reacted compound between cerium and steel leading to the isolation of anodic sites from cathodic sites. They still specify that the base property of FBE in regards to resistance to cathodic delamination was rather good. Ramenzanzadeh et al. [162] also studied a surface treatment based on neodymium compounds. They also obtained a better behavior when the resistance to salt spray and cathodic delamination were considered still due to enhancement of the adhesion between the steel substrate and FBE.

In addition to improve the adhesion between the steel substrate and FBE, modifications of FBE with different types of additives are studied to enhance FBE properties such as mechanical, conductive properties [152,163–166]. For example, carbon black can be added in FBE formulation [165]. When dispersed in FBE it can generate conductive or semi conductive coatings which can improve the coating barrier properties, investigated by EIS in 3% NaCl, due to the formation of dense carbon black network. Modification of FBE with acrylonitrile can also be quoted for the improvement of the coating flexibility [166]. It is indeed crucial as pipelines are not necessarily straight [151].

As similar for traditional solvent-borne coatings, the application parameters of FBE need to be carefully controlled. Incorrect curing temperature or time of curing can lead to adhesion failure between the FBE and the steel surface [167,168]. In addition to influence the flow viscosity, the surface wetting and the adhesion to the steel, the temperature of application can also impact the mechanical properties of the steel due to strain aging [169]. The high temperature of curing can also play a role in the internal stress of the coating and also lead to adhesion failure [154,170]. This is particularly true when a coating consisting in several layers is applied.

Several works have shown that FBE could present failure when used at high temperature due to its sensibility to water transport to the interface [171–173]. It is not uncommon to have high service temperature in offshore applications. Indeed, pipelines can carry hot fluid and be immerged in cold water. This created a thermal gradient. Payer et al. [171] showed that FBE was particularly sensible to negative thermal gradient where the steel is cooler than the coating. The service temperature can also impact the mechanical properties of the coating and lead to plasticization [154]. In addition, a high temperature of service can cause water to be absorbed more quickly and cause disbondment if surface contaminants are present [174].

As said, FBE coating can be designed as single coat, however, it is not unusual to associate it with layers of thermoplastic coating, such as polyethylene or polypropylene. These types of coating are called three-layer polyolefin systems (3PLO) which contain an FBE primer, an adhesive tie coat and a polyolefin outer layer [149]. These coatings generally exhibit better anticorrosive performance as single layer FBE because the polyolefin layer offers outstanding mechanical resistance to damage occurring during the pipe manufacture and installation [149].

6. Conclusions

The use of organic coatings as protection for metal surface is the still one of the popular methods when corrosion issue is considered. Organic coatings act as barrier against the corrosive species and can also provide additional protection through the use of pigments (sacrificial or inhibitive passivation). Constant changes in coating technology is carried out to meet either enhanced performance or to meet requirements due to environmental regulations, in particular the reduction of VOCs. To this end, products with high-solid contents, powder coatings or waterborne coatings with low amounts of organic solvents were developed. The challenge to coating suppliers is to find a solution with a more environmentally friendly profile without sacrificing the proven performance features of the traditional solvent-borne coatings. These new technologies are now well developed and challenge in terms of formulation and paint applications were found. As for traditional coatings, these types of low VOC paints still undergo continual change to improve their properties and meet the particular demands for the protection of steel in marine and offshore environments.

Even though, high solid and powder coatings technology seem to be well developed nowadays, in terms of production and application, they are not systematically used as replacement for traditional solvent-borne coatings in particular for heavy duty anticorrosive coatings. Indeed, this sector is more demanding on the coating, in terms of mechanical and physicochemical properties but also on the application and maintenance of the coating. One explanation for the lack of replacement of traditional solvent-borne coatings, could lie in the fact that end-users have less perspective in their long-term anticorrosive properties as for traditional solvent-borne. This is, however, less true for powder coatings where fusion bonded epoxy coatings are often used as anticorrosive solutions for pipeline protection. Indeed, the market for powder coating for corrosion protection of steel is expected to grow in future years. Another explanation, in the case of high-solid coatings, could be in the fact that the application of these types of coating is not as easy as for traditional coating and incorrect paint application can happen. A better surface preparation is also needed for high-solid and powder coatings.

In any case, even if these technologies (high-solid and powder coatings) are now well developed and known, as for every technology, constant improvement is researched and it will certainly allow them to gain more impact in the application in marine and offshore conditions.

Author Contributions: Writing—review and editing, K.P. and D.T.; supervision, D.T. All authors have read and agreed to the published version of the manuscript.

Funding: This research received no external funding.

Conflicts of Interest: The authors declare no conflict of interest.

References

1. Sørensen, P.A.; Kiil, S.; Dam-Johansen, K.; Weinell, C.E. Anticorrosive coatings: A review. *J. Coat. Technol. Res.* **2009**, *6*, 135–176. [CrossRef]
2. Bierwagen, G.P. Reflections on corrosion control by organic coatings. *Prog. Org. Coat.* **1996**, *28*, 43–48. [CrossRef]
3. Kumar, V.; Bhattacharya, A. Demand for Low-VOC Coatings Continues to Rise. 2020. Available online: https://www.pcimag.com/articles/107370-demand-for-low-voc-coatings-continues-to-rise (accessed on 23 September 2020).
4. Knudsen, O.Ø.; Forsgren, A. Weathering and Aging of paint. In *Corrosion Control Through Organic Coatings*, 2nd ed.; Philip, A., Schweitzer, P.E., Eds.; CRC Press: Boca Raton, FL, USA, 2017; Chapter 11; pp. 149–180.
5. Lyon, S.B.; Bingham, R.; Mills, D.J. Advances in corrosion protection by organic coatings: What we know and what we would like to know. *Prog. Org. Coat.* **2017**, *102*, 2–7. [CrossRef]
6. Mills, D.J.; Jamali, S. The best tests for anti-corrosive paints. And why: A personal viewpoint. *Prog. Org. Coat.* **2017**, *102*, 8–17. [CrossRef]
7. Khanna, A.S. Key issues in applying organic paint coatings. In *High Performance Organic Coatings*, 1st ed.; Khanna, A.S., Ed.; Woodhead Publishing Limited: Cambridge, UK, 2008; Chapter 1; pp. 1–27.

8. de Wit, J.H.M. Inorganic and Organic Coatings. In *Corrosion Mechanisms in Theory and Practice*, 1st ed.; Marcus, P., Oudar, J., Eds.; Dekker Inc.: New York, NY, USA, 1995; pp. 581–628.
9. Leidheiser, H. Corrosion of painted metals—A Review. *Corrosion* **1982**, *38*, 374–383. [CrossRef]
10. Mayne, J.E.O. How paints prevent corrosion. *Anti-Corros. Methods Mater.* **1954**, *1*, 286–290. [CrossRef]
11. Thomas, N.L. The barrier properties of paint coatings. *Prog. Org. Coat.* **1991**, *19*, 101–121. [CrossRef]
12. Kittelberger, W.W.; Elm, A.C. Diffusion of chloride through various paint systems. *Ind. Eng. Chem. Res.* **1952**, *44*, 326–329. [CrossRef]
13. Bacon, R.C.; Smith, J.J.; Rugg, F.M. Electrolytic resistance in evaluating protective merit of coatings on metals. *Ind. Eng. Chem. Res.* **1948**, *40*, 161–167. [CrossRef]
14. Kinsella, E.M.; Mayne, J.E.O. Ionic conduction in polymers films I. Influence of electrolyte resistance. *Br. Polym. J.* **1969**, *1*, 173–176. [CrossRef]
15. Mayne, J.E.O.; Scantelbury, J.D. Ionic conduction in polymer films II. Inhomogeneous structure of varnish films. *Br. Polym. J.* **1970**, *2*, 240–243. [CrossRef]
16. Croll, S.G. Electrolyte transport in polymer barrier coatings: Perspectives from other disciplines. *Prog. Org. Coat.* **2018**, *124*, 41–48. [CrossRef]
17. Sykes, J.M.; Whyte, E.P.; Pu, X.; Shaher Sahir, Z. Does "coating resistance" control corrosion? *Prog. Org. Coat.* **2017**, *102*, 82–87. [CrossRef]
18. Knudsen, O.Ø.; Forsgren, A. Corrosion-Protective Pigments. In *Corrosion Control Through Organic Coatings*, 2nd ed.; Philip, A., Schweitzer, P.E., Eds.; CRC Press: Boca Raton, FL, USA, 2017; Chapter 4; pp. 31–36.
19. Bieganska, B.; Zubielewicz, M.; Smieszek, E. Influence of barrier pigments on the performance of protective coatings. *Prog. Org. Coat.* **1988**, *16*, 219–229. [CrossRef]
20. Leidheiser, H.; Wang, W.; Ingetoft, L. The mechanism for the cathodic delamination of organic coatings from a metal surface. *Prog. Org. Coat.* **1983**, *11*, 19–40. [CrossRef]
21. Hare, C.H.; Wright, S. Anti-corrosive primers based on zinc flake. *J. Coat. Technol.* **1982**, *54*, 65–77.
22. Feliu, S.; Barajas, R.; Bastisdas, J.M.; Morcillo, M. Mechanism of cathodic protection of zinc-rich paints by electrochemical impedance spectroscopy. 1. Galvanic Stage. *J. Coat. Technol.* **1989**, *61*, 63–69.
23. Feliu, S.; Barajas, R.; Bastisdas, J.M.; Morcillo, M. Mechanism of cathodic protection of zinc-rich paints by electrochemical impedance spectroscopy. 2. Barrier Stage. *J. Coat. Technol.* **1989**, *61*, 71–76.
24. Kalendova, A. Effects of particle sizes and shapes of zinc metal on the properties of anticorrosive coatings. *Prog. Org. Coat.* **2003**, *46*, 324–332. [CrossRef]
25. Rammelt, U.; Reinhard, R. Characterization of active pigments in damage of organic coatings on steel by means of electrochemical impedance spectroscopy. *Prog. Org. Coat.* **1994**, *24*, 309–322. [CrossRef]
26. Romagnoli, R.; Vetere, V.F. Non-pollutant corrosion inhibitive pigments: Zinc phosphate, a review. *Corros. Rev.* **1995**, *13*, 45–64. [CrossRef]
27. Chromy, L.; Kaminska, E. Non-toxic anticorrosive pigments. *Prog. Org. Coat.* **1990**, *18*, 319–324. [CrossRef]
28. Deyà, M.; Vetere, V.F.; Romagnoli, R.; del Amo, B. Zinc tripolyphosphate: An anticorrosive pigment for paints. *Surf. Coat. Int. Part B* **2003**, *86*, 79–85. [CrossRef]
29. Pryor, M.J.; Cohen, M. The inhibition of the corrosion of iron by some anodic inhibitors. *J. Electrochem. Soc.* **1953**, *100*, 203–215. [CrossRef]
30. Leidheiser, H. Mechanism of corrosion inhibition with special attention to inhibitors in organic coatings. *J. Coat. Technol.* **1981**, *53*, 29–39.
31. Clay, M.F.; Cox, J.H. Chromate and phosphate pigments in anti-corrosive primers. *J. Oil Colour Chem.* **1973**, *56*, 13–16.
32. Hernandez, M.; Genesca, J.; Uruchurtu, J.; Galliano, F.; Landolt, D. Effect of inhibitive pigment zinc-aluminium-phosphate (ZAP) on the corrosion mechanisms of steel in waterborne coatings. *Prog. Org. Coat.* **2006**, *56*, 199–206. [CrossRef]
33. Zubielewicz, M.; Gnot, W. Mechanisms of non-toxic anticorrosive pigments in organic waterborne coatings. *Prog. Org. Coat.* **2004**, *49*, 358–371. [CrossRef]
34. Abu Ayana, Y.M.; El-Sawy, S.M.; Salah, S.H. Zinc-ferrite pigment for corrosion protection. *Anti-Corros. Methods Mater.* **1997**, *44*, 381–388. [CrossRef]
35. Beland, M. New non-toxic pigment performance profile equivalent to zinc chromate. *Am. Paint Coat. J.* **1991**, *76*, 43–50.

36. Funke, W. Problems and progress in organic coatings science and technology. *Prog. Org. Coat.* **1997**, *31*, 5–9. [CrossRef]
37. Funke, W. How organic coatings systems protect against corrosion. *Polym. Mater. Control Corros. Control.* **1986**, 200–228. [CrossRef]
38. McBain, J.W.; Hopkins, D.G. On adhesives and adhesive action. *J. Phys. Chem.* **1925**, *29*, 188–204. [CrossRef]
39. Sharpe, L.H.; Schonhorn, H. Surface energetics, adhesion and adhesive joints. *Adv. Chem. Ser.* **1964**, *8*, 189–201. [CrossRef]
40. Voyutski, S.S. Autohesion and adhesion of high polymers. In *Polymer Reviews*; Wiley: New York, NY, USA, 1963; Volume 4.
41. Buchan, S.; Rae, W.D. Chemical nature of the rubber to glass bond. *Trans. Inst. Rubber Ind.* **1946**, *20*, 205–216.
42. Bikerman, J.J. Causes of poor adhesion: Weak boundary layers. *Ind. Eng. Chem.* **1967**, *59*, 40–44. [CrossRef]
43. Baldan, A. Adhesion phenomena in bonded joints. *Int. J. Adhes. Adhes.* **2012**, *38*, 95–116. [CrossRef]
44. Fourche, G. An overview of the basic aspects of polymer adhesion. Part I: Fundamentals. *Polym. Eng. Sci.* **1995**, *35*, 957–967. [CrossRef]
45. Marsh, J.; Scantlebury, J.D.; Lyon, S.B. The effect of surface/primer treatments on the performance of alkyd coated steel. *Corros. Sci.* **2001**, *43*, 829–852. [CrossRef]
46. Leidheiser, H.; Funke, W. Water disbondment and wet adhesion of organic coatings on metals: A review and interpretation. *J. Oil Color Chem. Assoc.* **1987**, *70*, 121–132.
47. Greenfield, D.; Scantelbury, J.D. The protective action of organic coatings on steel: A review. *J. Corros. Sci. Eng.* **2000**, *3*, 5.
48. Funke, W. Blistering of paint films and filiform corrosion. *Prog. Org. Coat.* **1981**, *9*, 29–46. [CrossRef]
49. Pommersheim, J.M.; Nguyen, T. Prediction of blistering in coating systems. *ACS Symp. Ser.* **1998**, *689*, 137–150. [CrossRef]
50. De la Fuente, D.; Rohwerder, M. Fundamental investigation on the stability of the steel/coating interfaces contaminated by submicroscopy salt particles. *Prog. Org. Coat.* **2008**, *61*, 233–239. [CrossRef]
51. Morcillo, M. Soluble salts: Their effect on premature degradation of anticorrosive paints. *Prog. Org. Coat.* **1999**, *36*, 137–147. [CrossRef]
52. Khanna, A.S. Surface preparation for organic paint coatings. In *High Performance Organic Coatings*, 1st ed.; Khanna, A.S., Ed.; Woodhead Publishing Limited: Cambridge, UK, 2008; Chapter 2; pp. 27–41.
53. ISO Standard. *ISO 8501 Preparation of Steel Substrates before Application of Paints and Related Products—Visual Assessment of Surface Cleanliness*; ISO Standard: Brussels, Belgium, 2007.
54. Le Calvé, P.; Pautasso, J.-P.; Le Bozec, N. Characterizing Surfaces after UHP Waterjetting in New Ship Construction. 2011. Available online: https://www.paintsquare.com/store/assets/JPCL_uhp_ebook1.pdf#page=38 (accessed on 23 September 2020).
55. Le Calvé, P.; Pautasso, J.-P.; Le Bozec, N. Characterizing Surfaces after UHP Waterjetting on Shop-Primed Steel in New Ship Construction, Part 3. 2012. Available online: https://search.proquest.com/openview/e9c43a2ce5adcfc5d99d620eb9dde947/1?pq-origsite=gscholar&cbl=36623 (accessed on 23 September 2020).
56. Knudsen, O.Ø.; Forsgren, A. Blast cleaning and other heavy surface pretreatments. In *Corrosion Control Through Organic Coatings*, 2nd ed.; Philip, A., Schweitzer, P.E., Eds.; CRC Press: Boca Raton, FL, USA, 2017; Chapter 7; pp. 89–108.
57. Momber, A.W.; Greverath, W.D. Surface preparation standards for steel substrates—A critical review. *J. Prot. Coat. Linings* **2004**, *21*, 48–52.
58. Momber, A.W.; Koller, S.; Dittmers, H.J. Effects of surface preparation methods on adhesion of organic coatings to steel substrates. *J. Prot. Coat. Linings* **2004**, *21*, 44–50.
59. Hare, C. Internal stress-related coating system failures. *J. Prot. Coat. Linings* **1996**, *13*, 99–113.
60. Sato, K. The internal stress of coating films. *Prog. Org. Coat.* **1980**, *8*, 143–160. [CrossRef]
61. Clyne, T.W. Residual stresses in surface coatings and their effects on interfacial debonding. *Key Eng. Mater.* **1996**, *5*, 401–406. [CrossRef]
62. Piens, M.; de Deurwaerder, H. Effect of coating stress on adherence and on corrosion protection. *Prog. Org. Coat.* **2001**, *43*, 18–24. [CrossRef]
63. Negele, O.; Funke, W. Internal stress and wet adhesion of organic coatings. *Prog. Org. Coat.* **1996**, *28*, 285–289. [CrossRef]

64. Nazarov, A.; Thierry, D. Application of Scanning Kelvin Probe in the study of protective paints. *Front. Mater.* **2019**, *6*, 1–17. [CrossRef]
65. Sharman, C.F. Filiform underfilm corrosion of lacquered steel surfaces. *Nature* **1944**, *153*, 621–622. [CrossRef]
66. Castle, J.E.; Watts, J.F. Cathodic disbondment of well characterized steel/coating interfaces. In *Corrosion Control by Organic Coatings*; Leidhieser, H., Jr., Ed.; NACE International: Houston, TX, USA, 1981; p. 78.
67. Hammond, J.S.; Holubka, J.W.; de Vries, J.E.; Dickie, R.A. The application of X-ray photoelectron spectroscopy to a study of interfacial composition in corrosion-induced paint de-adhesion. *Corros. Sci.* **1981**, *21*, 239–253. [CrossRef]
68. Sorensen, P.A.; Dam-Johansen, K.; Weinell, C.E.; Kiil, S. Cathodic delamination of seawater-immersed anticorrosive coatings: Mapping of parameters affecting the rate. *Prog. Org. Coat.* **2010**, *68*, 283–292. [CrossRef]
69. Sorensen, P.A.; Dam-Johansen, K.; Weinell, C.E.; Kiil, S. Cathodic delamination: Quantification of ionic transport rates along coating-steel interfaces. *Prog. Org. Coat.* **2010**, *68*, 70–78. [CrossRef]
70. Nazarov, A.; Thierry, D. Mechanism of the corrosion exfiolation of a polymer coating from a carbon steel. *Prot. Met. Phys. Chem. Surf.* **2009**, *45*, 735–745. [CrossRef]
71. Nazarov, A.; Thierry, D.; Le Calvé, P.; Pautasso, J.-P. Scanning Kelvin Probe investigation of corrosion under thick marine paint systems applied on carbon steel. *Corrosion* **2012**, *68*, 720–729. [CrossRef]
72. Steel, G.D. Filiform corrosion on architectural aluminium—A review. *Anti-Corros. Methods Mater.* **1994**, *41*, 8–12. [CrossRef]
73. Slabaugh, W.H.; Hutchins, L.L.; Dejager, W.; Hoover, S.E. Filiform corrosion of aluminium. *J. Paint Technol.* **1972**, *44*, 76–80.
74. Olsen, H.; Nisancioglu, K. Filiform corrosion of aluminium sheet. I. Corrosion behaviour of painted steel. *Corros. Sci.* **1998**, *40*, 1179–1194. [CrossRef]
75. de Wit, J.H.W.; Mol, J.M.C. Organic coatings for marine and shipping applications. In *High Performance Organic Coatings*, 1st ed.; Khanna, A.S., Ed.; Woodhead Publishing Limited: Cambridge, UK, 2008; Chapter 16; pp. 337–388.
76. Ngyuen, T.N.; Hubbard, J.B.; McFadden, G.B. A mathematical model for the cathodic blistering of organic coatings on steel immersed in electrolytes. *J. Coat. Technol.* **1991**, *63*, 43–52.
77. Funke, W. Toward a unified view of the mechanism responsible for paint defects by metallic corrosion. *Ind. Eng. Chem. Prod. Res. Dev.* **1985**, *24*, 343–347. [CrossRef]
78. Hare, C.H. Mechanisms of photolytically induced degradation. *J. Prot. Coat. Linings* **2000**, *17*, 73–86.
79. Galant, C.; Fayolle, B.; Kuntz, M.; Verdu, J. Thermal and radio-oxidation of epoxy coating. *Prog. Org. Coat.* **2010**, *69*, 322–329. [CrossRef]
80. Pathak, S.S.; Khanna, A.S. Waterborne coatings for corrosion protection. In *High Performance Organic Coatings*, 1st ed.; Khanna, A.S., Ed.; Woodhead Publishing Limited: Cambridge, UK, 2008; Chapter 13; pp. 247–286.
81. Hill, L.W. Design consideration for high solids coatings. *Prog. Org. Coat.* **1982**, *10*, 55–89. [CrossRef]
82. Weiss, K.D. Paint and coatings: A mature industry in transition. *Prog. Polym. Sci.* **1997**, *22*, 203–245. [CrossRef]
83. Wu, S. Rheology of high solid coatings I. Analysis of Sagging and Slumping. *J. Appl. Polym. Sci.* **1978**, *22*, 2769–2782. [CrossRef]
84. Jones, F.N.; Nichols, M.E.; Pappas, S.P. Solventborne and high solids coatings. In *Organic Coatings: Science and Technology*, 4th ed.; Wiley, J., Ed.; John Wiley and Sons: Hoboken, NJ, USA, 2017; Chapter 25; pp. 357–365.
85. Bauer, D.R.; Budde, G.F. Cross-linking chemistry and network structure in high solids acrylic-melamine coatings. *Ind. Eng. Chem. Process. Des. Dev.* **1981**, *20*, 674–679. [CrossRef]
86. Takahashi, M. Recent advances in high solids coatings. *Polym. Plast. Technol. Eng.* **1980**, *15*, 1–49. [CrossRef]
87. Knudsen, O.Ø.; Forsgren, A. Powder Coatings. In *Corrosion Control through Organic Coatings*, 2nd ed.; Philip, A., Schweitzer, P.E., Eds.; CRC Press: Boca Raton, FL, USA, 2017; Chapter 6; pp. 71–87.
88. Du, Z.; Wen, S.; Wang, J.; Yin, C.; Yu, D.; Luo, J. The review of powder coatings. *J. Mater. Sci. Chem. Eng.* **2016**, *4*, 54–59. [CrossRef]
89. Kroschwitz, J.J.; Mark, H.F. Coatings methods, Powder technology. In *Encyclopedia of Polymer Science and Technology*, 4th ed.; Mark, H.F., Ed.; John Wiley and Sons: Hoboken, NJ, USA, 2004; Volume 5, pp. 570–603.
90. Jones, F.N.; Nichols, M.E.; Pappas, S.P. Powder coatings. In *Organic Coatings: Science and Technology*, 4th ed.; Wiley, J., Ed.; John Wiley and Sons: Hoboken, NJ, USA, 2017; Chapter 28; pp. 385–402.

91. Spyrou, E. *Powder Coatings Chemistry and Technology*, 3rd ed.; Vincentz Network GmbH: Hanover, Germany, 2012.
92. Bailey, A.G. The science and technology of electrostatic powder spraying, transport and coating. *J. Electrost.* **1998**, *45*, 85–120. [CrossRef]
93. Boncza-Tomaszewski, Z.; Penczek, P. Tribo charging powder coatings. *Macromol. Symp.* **2002**, *187*, 417–426. [CrossRef]
94. Karidkar, S.; Mali, R. Optimization of powder spray process parameters using Taguchi methodology. *Adv. Intell. Syst. Res.* **2017**, *137*, 71–76. [CrossRef]
95. Uhlmann, P.; Grundke, K. Influence of additives on interfacial phenomena during film formation of powder coatings. *J. Coat. Technol.* **2001**, *73*, 59–65. [CrossRef]
96. Barletta, M.; Lusvarghi, L.; Pighetti Mantini, F.; Rubino, G. Epoxy-based thermosetting powder coatings: Surface appearance, scratch adhesion and wear resistance. *Surf. Coat. Technol.* **2007**, *201*, 7449–7504. [CrossRef]
97. Mafi, R.; Mirabedini, S.M.; Naderi, R.; Attar, M.M. Effect of curing characterization on the corrosion performance of polyester and polyester/epoxy powder coatings. *Corros. Sci.* **2008**, *50*, 3280–3286. [CrossRef]
98. Banerje, S.; Mazumder, M.K. Adhesion of charged powders to metal surface in the powder coating process. *IEEE Trans. Ind. Appl.* **1996**, *32*, 1243–1249. [CrossRef]
99. ISO Standard. *ISO 12944-5: Paints and Varnishes—Corrosion Protection of Steel Structures by Protective Paints—Part 5: Protective Paint Systems*; ISO Standard: Brussels, Belgium, 2019.
100. Knudsen, O.Ø.; Forsgren, A. Generic types of anticorrosion coatings. In *Corrosion Control through Organic Coatings*, 2nd ed.; Philip, A., Schweitzer, P.E., Eds.; CRC Press: Boca Raton, FL, USA, 2017; Chapter 3; pp. 11–30.
101. Kanitkar, S. High performance epoxies and solvent-less epoxies for corrosion protection. In *High Performance Organic Coatings*, 1st ed.; Khanna, A.S., Ed.; Woodhead Publishing Limited: Cambridge, UK, 2008; Chapter 8; pp. 145–164.
102. Banfield, T.A. The protective aspects of marine paints. *Prog. Org. Coat.* **1979**, *7*, 253–277. [CrossRef]
103. Heaner, W.L.; Aguirre-Vargas, F.; Ge, S.; Xu, N.; Krishnan Karunakaran, K.; Weishuhn, J. *Tunning Toughness and Flexibility in Liquid Applied High Solids Epoxy Coatings*; Corpus ID: 102500379; The Dow Chemical Company: Midland, TX, USA, 2016.
104. Saunders, T.F.; Lévy, M.F.; Serino, F.F. Mechanisms of the tertiary amine-catalyzed dicyandiamide cure of epoxy resins. *J. Polym. Sci. Part A* **1967**, *5*, 1609–1617. [CrossRef]
105. Zahir, S.A. The mechanism of cure epoxide resins by cyanamide and dicyandiamide. *Adv. Org. Coat. Sci. Technol.* **1982**, *4*, 82–102.
106. Gilbert, M.D.; Schneider, N.S. Mechanism of the dicyandiamide/epoxide reaction. *Macromolecules* **1991**, *24*, 360–369. [CrossRef]
107. Gaukler, J.C. Oligomer formation in epoxy-dicyandiamide systems. *J. Adhes.* **2012**, *88*, 720–750. [CrossRef]
108. Dickerson, J.G. FBE evolves to meet industry need for pipeline protection. *Pipeline Gas Ind.* **2001**, *84*, 64–74.
109. Kehr, A.; Enos, D.G. FBE, a foundation for pipeline corrosion coatings. In Proceedings of the NACE International's CORROSION Conference, Orlando, FL, USA, 26–31 March 2000; p. 00757.
110. Kehr, A.; Dabiri, M.; Hislop, R. Dual-Layer Fusion Bonded Epoxy (FBE) Coatings Protect Pipelines. In Proceedings of the INDOPIPE, Jakarta, Indonesia, 30 May–1 June 2006.
111. Chattopadhyay, D.K.; Raju, K.V.S.N. Structural engineering of polyurethane coatings for high performance applications. *Prog. Polym. Mater. Sci.* **2007**, *32*, 352–418. [CrossRef]
112. Somarathna, H.M.C.C.; Raman, S.N.; Mohotti, D.; Mutalib, A.A.; Badri, K.H. The use of polyurethane for structural and infrastructural engineering applications: A state-of-the-art review. *Constr. Build. Mater.* **2018**, *190*, 995–1104. [CrossRef]
113. Gao, T.; He, Z.; Hihara, L.H.; Mehr, H.S.; Soucek, M.D. Outdoor exposure and accelerated weathering of polyurethane/polysiloxane hybrid coatings. *Prog. Org. Coat.* **2019**, *130*, 44–55. [CrossRef]
114. Keene, L.T.; Halada, G.P.; Clayton, C.R. Failure of navy coatings systems 1: Chemical depth profiling of artificially and naturally weathered high-solids aliphatic poly(ester-urethane) military coating systems. *Prog. Org. Coat.* **2005**, *52*, 173–186. [CrossRef]
115. ASTM Standard D16-00. *Technology for Paint, Related Coatings, Materials and Applications*; ASTM International: West Conshohocken, PA, USA, 2014.

116. Potter, T.A.; Schmelzer, H.G.; Baker, R.D. High-solids coatings based on polyurethane chemistry. *Prog. Org. Coat.* **1984**, *12*, 321–328. [CrossRef]
117. Guan, S.W. 100% solids polyurethane and polyurea coatings technology. *Coat. World* **2003**, 49–58.
118. Renz, H.; Bruchmann, B. Pathways targeting solvent-free PUR coatings. *Prog. Org. Coat.* **2001**, *43*, 32–40. [CrossRef]
119. Mannari, V.M.; Massingill, J.L. Two-component high-solid polyurethane coatings systems based on soy polyols. *J. Coat. Technol. Res.* **2006**, *3*, 151–157. [CrossRef]
120. Misev, T.A.; van der Linde, R. Powder coatings technology: New developments at the turn of the century. *Prog. Org. Coat.* **1998**, *34*, 160–168. [CrossRef]
121. Kaesler, K.H. Polysiloxane coatings for corrosion protection. In *High performance Organic Coatings*, 1st ed.; Khanna, A.S., Ed.; Woodhead Publishing Limited: Cambridge, UK, 2008; Chapter 12; pp. 225–246.
122. Chrusciel, J.J.; Lesniak, E. Modification of epoxy resins with functional silanes, polysiloxanes, silsesquioxanes, silica and silicates. *Prog. Polym. Sci.* **2015**, *41*, 67–121. [CrossRef]
123. Byczynski, L.; Dutkiewicz, M.; Maciejewski, H. Synthesis and properties of high-solids hybrid materials obtained from epoxy functional urethanes and siloxanes. *Prog. Org. Coat.* **2015**, *84*, 59–69. [CrossRef]
124. Gommans, J.H.P. Polysiloxane Coatings—A mature technology. In Proceedings of the Corrosion & Prevention Conference, Auckland, New Zealand, 1 January 2000.
125. Parashar, G.; Srivastava, D.; Kumar, P. Ethyl silicate binders for high performance coatings. *Prog. Org. Coat.* **2001**, *42*, 1–14. [CrossRef]
126. ISO Standard. *ISO 12944-2: Paints and Varnishes—Corrosion Protection of Steel Structures by Protective Paints—Part 2: Environment Classification*; ISO Standard: Brussels, Belgium, 2019.
127. Thierry, D.; Persson, D.; Le Bozec, N. Atmospheric corrosion of zinc and zinc alloyed coated steel. In *Encyclopedia of Interfacial Chemistry: Surface Science and Electrochemistry*, 1st ed.; Wandelt, K., Ed.; Elsevier: Amsterdam, The Netherlands, 2018; pp. 55–78. [CrossRef]
128. Olivier, M.-G.; Romano, A.P.; Vandermeiers, C.; Mathieu, X.; Boelman, M. Influence of the stress generated during an ageing cycle on the barrier properties of catophoretic coatings. *Prog. Org. Coat.* **2008**, *63*, 323–329. [CrossRef]
129. Drodten, P. Marine atmosphere, splash, tidal and immersion zones. In *Corrosion Handbook Online*, 1st ed.; Kreysa, G., Schutze, M., Eds.; Dechema: Frankfurt, Germany, 2008; pp. 263–322. [CrossRef]
130. Almeide, E.; Diamantino, T.C.; de Sousa, O. Marine paints: The particular case of antifouling paints. *Prog. Org. Coat.* **2007**, *59*, 2–20. [CrossRef]
131. Chambers, L.D.; Stokes, K.R.; Walsh, F.C.; Wood, R.J.K. Modern approaches to marine antifouling coatings. *Surf. Coat. Technol.* **2006**, *201*, 3642–3652. [CrossRef]
132. Detty, M.; Ciriminna, R.; Bright, F.V.; Pagliaro, M. Environmentally benign sol-gel antifouling and foul-releasing coatings. *Acc. Chem. Res.* **2014**, *47*, 678–687. [CrossRef]
133. Buskens, P.; Wouters, M.; Rentrop, C.; Vroon, Z. A brief review of environmentally benign antifouling and foul-release coatings for marine applications. *J. Coat. Technol. Res.* **2013**, *10*, 29–36. [CrossRef]
134. Shi, W.; Lyon, S.B. Investigation using localized SVET into protection at defects in epoxy coated mild steel under intermittent cathodic protection simulating inter-tidal and splash zones. *Prog. Org. Coat.* **2017**, *102*, 66–70. [CrossRef]
135. Yang, X.; Zhang, T.; Wang, H.; Hou, B. A new solvent-free super high build epoxy coating evaluated by marine corrosion simulation apparatus. *Mater. Corros.* **2012**, *63*, 328–332. [CrossRef]
136. Zubielewicz, M.; Krolikowska, A. The influence of ageing of epoxy coatings on adhesion of polyurethane topcoats and protective properties of coating systems. *Prog. Org. Coat.* **2009**, *66*, 129–136. [CrossRef]
137. Le Bozec, N.; Thierry, D.; Le Calvé, P.; Favennec, C.; Pautasso, J.-P.; Hubert, C. Performance of marine and offshore paints systems: Correlation of accelerated corrosion tests and field exposure on operating ships. *Mater. Corros.* **2013**, *66*, 215–225. [CrossRef]
138. Le Bozec, N.; Thierry, D.; Pélissier, K. A new accelerated corrosion test for marine paint systems used for ship's topsides and superstructures. *Mater. Corros.* **2018**, *69*, 447–459. [CrossRef]
139. Rezaei, F.; Sharif, F.; Sarabi, A.A.; Kasiriha, S.M.; Rahmanian, M.; Akbarinezhad, E. Experimental evaluation of high solid polyurethane coating in the presence of salt at high temperature. *Mater. Corros.* **2010**, *61*, 681–688. [CrossRef]

140. Diaz, I.; Chico, B.; de la Fuente, D.; Simancas, J.; Vega, J.M.; Morcillo, M. Corrosion resistance of new epoxy-siloxane hybrid coatings. A laboratory study. *Prog. Org. Coat.* **2010**, *69*, 278–286. [CrossRef]
141. ISO Standard. *ISO 12944-6: Paints and Varnishes—Corrosion Protection of Steel Structures by Protective Paints—Part 6: Laboratory Performance Test Methods*; ISO Standard: Brussels, Belgium, 2019.
142. Xu, S.; Yang, J.; Ma, C. Analysis on causes of polysiloxanes surface coating peeling on a special ship. *Key Eng. Mater.* **2019**, *815*, 37–41. [CrossRef]
143. Knudsen, O.O. Review of coating failure incidents on the Norwegian continental shelf since the introduction of NORSOK-501. In Proceedings of the NACE International's CORROSION Conference, Orlando, FL, USA, 17–21 March 2013.
144. Husain, A.; Al Madaj, M. Optimization of AC impedance test cell for accelerated evaluation of marine industrial coating in Kuwait. *Desalin. Water Treat.* **2013**, *51*, 1980–1988. [CrossRef]
145. Shreepathi, S. Physicochemical parameters influencing the testing of cathodic delamination resistance of high build pigmented epoxy coating. *Prog. Org. Coat.* **2016**, *90*, 438–447. [CrossRef]
146. Santos, D.; Brites, C.; Costa, M.R.; Santos, M.T. Performance of paint systems with polyurethane topcoats, proposed for atmospheres with very high corrosvity category. *Prog. Org. Coat.* **2005**, *54*, 344–352. [CrossRef]
147. Sanchez-Amaya, J.M.; Osuna, R.M.; Bethencourt, M.; Botana, F.J. Monitoring the degradation of a high solids epoxy coating by means of EIS and ENS. *Prog. Org. Coat.* **2007**, *60*, 248–254. [CrossRef]
148. Guan, S.W.; Liu, D.; Moreno, M.; Garneau, R. 100% solids rigid polyurethane coatings technology for corrosion protection of ballast tanks. In Proceedings of the NACE International's CORROSION Conference, New Orleans, LA, USA, 28 March–1 April 2004.
149. Fathabadi, H.E. Investigation of fusion bonded epoxy (FBE) as a protective coating for pipes. In Proceedings of the 3rd International Color and Coating Congress, Teheran, Iran, 16 November 2009.
150. Khanna, A.S. Application for high-performance organic coatings. In *High Performance Organic Coatings*, 1st ed.; Khanna, A.S., Ed.; Woodhead Publishing Limited: Cambridge, UK, 2008; Chapter 14; pp. 287–307.
151. Norsworthy, R. Selection and use of coatings for underground or submersion service. In *Uhlig's Corrosion Handbook*, 3rd ed.; Winston Revie, R., Ed.; John Wiley and Sons: Hoboken, NJ, USA, 2011; Chapter 68; pp. 985–1000. [CrossRef]
152. Huttunen-Saarivirta, E.; Vagarov, G.V.; Yudin, V.E.; Vuorinen, J. Characterization and corrosion protection properties of epoxy powder coatings containing nanoclays. *Prog. Org. Coat.* **2013**, *76*, 757–767. [CrossRef]
153. Boerschel, V. New developments of mid-Tg FBE powder coatings to meet the requirements of pipe coaters and pipeline owners. In Proceedings of the NACE International's CORROSION Conference, San Antonio, TX, USA, 14–18 March 2010. Paper 10012.
154. Mobin, M.; Malik, A.U.; Andijani, I.N.; Al-Muaili, F.; Al-Hajri, M.; Ozair, G.; Mohammad, N.M.K. Performance evaluation of some fusion-bonded epoxy coatings under water transmission line conditions. *Prog. Org. Coat.* **2008**, *62*, 369–375. [CrossRef]
155. Cetiner, M.; Abes, J.; Slingh, P.; Gilroy-Scott, A. UV degradation of fusion bonded epoxy coating in stockpiled pipes. In Proceedings of the International Pipeline Conference, Calgary, AB, Canada, 1–5 October 2000.
156. Ramezanzadesh, B.; Rostami, M. The effect of cerium-based conversion treatment on the cathodic delamination and corrosion protection performance of carbon steel fusion bonded epoxy coatings. *Appl. Surf. Sci.* **2017**, *392*, 1004–1016. [CrossRef]
157. Papavinasam, S. Mitigation-External Corrosion. In *Corrosion Control in the Oil and Gas Industry*, 1st ed.; Gulf Professional Publishing: London, UK, 2014; Chapter 9; pp. 529–620. [CrossRef]
158. Saliba, P.A.; Mansur, A.A.; Santos, D.B.; Mansur, H.S. Fusion bonded epoxy composite coatings on chemically functionalized API steel surfaces for potential deep-water petroleum exploration. *Appl. Adhes. Sci.* **2015**, *3*, 1–22. [CrossRef]
159. Rodriguez, R.E.; Trautman, B.L.; Payer, J.H. Influencing factors in cathodic disbondment of fusion bonded epoxy coatings. In Proceedings of the NACE International's CORROSION Conference, Orlando, FL, USA, 26–31 March 2000.
160. Ramezanzadesh, M.; Bahlakeh, G.; Ramezanzadesh, B.; Rostani, M. Mild steel surface eco-friendly treatment by neodymium-based nanofilm for fusion bonded epoxy coating anti-corrosion/adhesion properties enhancement in simulated seawater. *J. Ind. Eng. Chem.* **2019**, *72*, 474–490. [CrossRef]
161. Sabadi, M.; Terryn, H.; Mol, J.M.C. The influence of a Zr-based conversion treatment on interfacial bonding strength and stability of epoxy coated carbon steel. *Prog. Org. Coat.* **2017**, *105*, 29–36. [CrossRef]

162. Tchoquesi Diodjo, M.R.; Belec, L.; Aragon, E.; Joliff, Y.; Lanarde, L.; Perrin, F.-X. Silane coupling agent for attracting FBE to steel. *Acs Appl. Mater. Interfaces* **2013**, *5*, 6751–6761. [CrossRef] [PubMed]
163. Radhakrishnan, S.; Sonaware, N.; Siju, C.R. Epoxy powder coatings containing polyaniline for enhanced corrosion protection. *Prog. Org. Coat.* **2009**, *64*, 383–386. [CrossRef]
164. Ruhi, G.; Bhandani, H.; Dhawan, S.K. Designing of corrosion resistant epoxy coatings embedded with polypyrrole/SiO_2 composite. *Prog. Org. Coat.* **2014**, *77*, 1484–1498. [CrossRef]
165. Wei, Y.H.; Zhang, L.X.; Ke, W. Evaluation of corrosion protection of carbon black filled FBE coatings on mild steel during exposure to a quiescent 3% NaCl. *Corros. Sci.* **2007**, *49*, 287–302. [CrossRef]
166. Jingcheng, L.; Xiuli, J.; Shengwen, Z.; Ren, L.; Xiaya, L. Preparation and characterization of carboxyl terminated poly(butadiene-co-acrylonitrile) epoxy resin prepolymers for FBE powder coatings. *J. Wuhan Univ. Technol. Mater. Sci. Ed.* **2012**, *27*, 694–701. [CrossRef]
167. Keyvani, B.; Sharif, A.; Khorassani, M. Determination of Arrhenius equation constants for curing reaction kinetics of fusion bonded epoxy. *Org. Chem. J.* **2010**, *1*, 15–19.
168. Chang, B.T.A.; Sue, H.-J.; Jiang, H.; Browing, B.; Wong, D.; Pham, H.; Guo, S.; Kehr, A.; Mallozzi, M.; Snider, W.; et al. Integrity of 3LPE pipeline coatings: Residual stresses and adhesion degradation. In Proceedings of the 7th International Pipeline Conference, Calgary, AB, Canada, 29 September–3 October 2008; p. 64048. [CrossRef]
169. Jadoon, A.N.K.; Thompson, I. Fusion bonded epoxy mainline and field joint coatings performance from the X100 field trial—A case study. *Int. J. Press. Vessel. Pip.* **2012**, *92*, 48–55. [CrossRef]
170. Joliff, Y.; Belec, L.; Aragon, E. Influence of the thickness of pipeline coating on internal stresses during the manufacture process by finite element analysis. *Comput. Mater. Sci.* **2013**, *68*, 342–349. [CrossRef]
171. Payer, J.H.; Moore, D.P.; Lee Magnon, J. Performance testing of fusion bonded epoxy coatings. In Proceedings of the NACE International's CORROSION Conference, Orlando, FL, USA, 26–31 March 2000.
172. Al Borno, A.; Chen, X.; Dhoke, S.K. Effect of high temperature sodium hydroxide immersion on FBE coating. *Int. J. Corros.* **2015**, *903478*, 1–7. [CrossRef]
173. Legghe, E.; Aragon, E.; Belec, L.; Margaillan, A.; Melot, D. Correlation between water diffusion and adhesion loss study of an epoxy primer steel. *Prog. Org. Coat.* **2009**, *66*, 276–280. [CrossRef]
174. Sauvant-Moret, V.; Kittel, J.; Melot, D.; Roche, M. Three layers polyolefin coatings: How the FBE primer properties govern the long-term adhesion. In Proceedings of the 17th International Conference on Pipeline Protection, Edinburgh, UK, 17–19 October 2007.

coatings

MDPI

Review

Uses of Scanning Electrochemical Microscopy (SECM) for the Characterization with Spatial and Chemical Resolution of Thin Surface Layers and Coating Systems Applied on Metals: A Review

Juan J. Santana [1,*], Javier Izquierdo [2,3] and Ricardo M. Souto [2,3,*]

1 Department of Process Engineering, University of Las Palmas de Gran Canaria, 35017 Las Palmas de Gran Canaria, Gran Canaria, Spain
2 Department of Chemistry, Universidad de La Laguna, 38200 La Laguna, Tenerife, Spain; jizquier@ull.edu.es
3 Institute of Material Science and Nanotechnology, Universidad de La Laguna, 38200 La Laguna, Tenerife, Spain
* Correspondence: juan.santana@ulpgc.es (J.J.S.); rsouto@ull.es (R.M.S.)

Abstract: Scanning Electrochemical Microscopy (SECM) is increasingly used in the study and characterization of thin surface films as well as organic and inorganic coatings applied on metals for the collection of spatially- and chemically-resolved information on the localized reactions related to material degradation processes. The movement of a microelectrode (ME) in close proximity to the interface under study allows the application of various experimental procedures that can be classified into amperometric and potentiometric operations depending on either sensing faradaic currents or concentration distributions resulting from the corrosion process. Quantitative analysis can be performed using the ME signal, thus revealing different sample properties and/or the influence of the environment and experimental variables that can be observed on different length scales. In this way, identification of the earlier stages for localized corrosion initiation, the adsorption and formation of inhibitor layers, monitoring of water and specific ions uptake by intact polymeric coatings applied on metals for corrosion protection as well as lixiviation, and detection of coating swelling—which constitutes the earlier stages of blistering—have been successfully achieved. Unfortunately, despite these successful applications of SECM for the characterization of surface layers and coating systems applied on metallic materials, we often find in the scientific literature insufficient or even inadequate description of experimental conditions related to the reliability and reproducibility of SECM data for validation. This review focuses specifically on these features as a continuation of a previous review describing the applications of SECM in this field.

Keywords: scanning electrochemical microscopy; corrosion protection; coating degradation; corrosion inhibitor films; electrochemical activity; microelectrode

Citation: Santana, J.J.; Izquierdo, J.; Souto, R.M. Uses of Scanning Electrochemical Microscopy (SECM) for the Characterization with Spatial and Chemical Resolution of Thin Surface Layers and Coating Systems Applied on Metals: A Review. *Coatings* **2022**, *12*, 637. https://doi.org/10.3390/coatings12050637

Academic Editor: Aomar Hadjadj

Received: 4 April 2022
Accepted: 1 May 2022
Published: 5 May 2022

Publisher's Note: MDPI stays neutral with regard to jurisdictional claims in published maps and institutional affiliations.

1. Introduction

Corrosion involves the destructive oxidation of metals and non-metallic materials, which causes degradation of their function as a result of exposure of the materials to environments that are aggressive to them. Most of the materials used in our society require a contribution of energy for their extraction and industrial production in the desired chemical state and form. Therefore, there is a great thermodynamic tendency for these materials to return to their original, more stable state—hence corrosion is ultimately an unavoidable process. However, the use of proper corrosion control methods significantly slows down the rate at which the corrosive phenomenon occurs, making it possible to increase the useful life of the material, thereby reducing the impact on the environment by a lower consumption of materials, namely by reducing the rate of replacement, as well as by the lower contamination due to degradation products.

The degradation reactions that occur in metallic materials exposed to the atmosphere or to an aqueous medium have their electrochemical origins in common. That is to say, they are produced by the development of electrochemical microcells on the surface of the corroding material, with dimensions typically within the micrometric and even submicrometric range in their beginnings. Thus, the anodic half-reaction (namely the oxidation of the metal) and the cathodic half-reaction (i.e., the oxygen reduction reaction in most aqueous environments) occur at different locations on the surface, forming highly localized microcells. As a result, several surface phenomena emerge, such as the formation of passive layers, local electric fields are generated, changes in the surface conductivity, or variations in the rate of electron transfer reactions, which, along with the possible coupling of homogeneous-phase reactions in solution, generation of local electric fields, changes in pH, etc., ultimately account for the degradation pathways, as shown in Figure 1. As most of the knowledge of corrosion mechanisms has been gathered using conventional electrochemical techniques that are surface averaging methods, little or no information is currently available on reactivity at sites of corrosion initiation or at small defects in surface layers and films. This situation is a major drawback for the development of effective corrosion protection technologies, which often can only attempt to minimise the extent of corrosion once it has started, as the actual mechanisms related to its initiation remain mostly unknown.

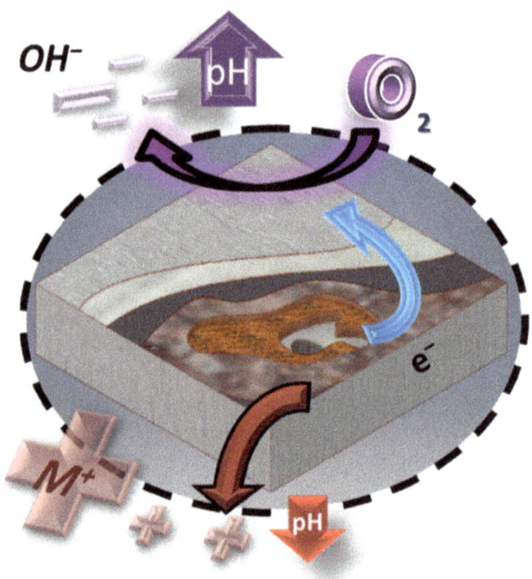

Figure 1. Sketch depicting the formation of localized microcells, the subsequent flow of ionic species in the electrolyte, and the pH changes associated with the onset of corrosion of a metal immersed in an alkaline or neutral aqueous medium.

The most effective and widely employed anticorrosion protection method consists of the application of organic coatings on metals. Organic coatings mainly provide a physical barrier against the access of water and ions to the metal surface, preventing the onset of electrochemical reactions there. However, there is no completely impermeable coating to these chemical species and therefore a certain electrochemical activity will find its origin in small defects, which are invariably present, but more notably in larger defects which are produced by the action of the environment (e.g., scratches, particle impacts, etc.) or on the cut edges of coated metals [1]. The addition of corrosion inhibitors could prolong the life of the coated material by inhibiting corrosive processes at these defects and cut edges, provided a steady supply of them is maintained when added to the coating as pigments.

This supply is not sustained in a controlled manner in typical anticorrosion coating formulation, so they are released by the coating continuously, eventually ceasing to be available for the protection of the material, and polluting the environment. A better concept is that the coatings could be functionalized for the release of the inhibitor only when corrosion has started, in order to stop the degradation processes and to heal the defect formed [2–4]. Such "smart" or self-repairing coatings, modified with additives and pigments specially designed for the mitigation of corrosive activity, would not release large amounts of chemicals into the environment, while providing more effective corrosion protection.

Although electrochemical techniques provide powerful tools to study interfacial reactions, especially corrosion processes, the conventional methods lack spatial resolution and provide limited information about electrochemical behaviour at sites of corrosion initiation or defects and cut edges. In fact, processes occurring at these sites are highly localized and heterogeneously distributed throughout the system (cf. Figure 1), and methods with spatial resolution that can acquire real-time data are needed to obtain relevant information about the underlying reaction mechanisms. In the last three decades, significant progress has been made in the knowledge and analytical monitoring of the origins of corrosion by means of electrochemical measurements at micrometric and sub-micrometric scale. Some methods have been devised to considerably reduce the number of degradation events that can occur simultaneously on a given substrate, even resulting in the electrochemical signal measured in the system coming from distinct events. This is achieved by miniaturizing the measurement cell [5] or the sample under study using individual microelectrodes [6] and microelectrode arrays [7]. The measured signal always represents an average of all events occurring on the exposed substrate, but transients related to individual breakdown events or localized sites can be distinguished above the background signal. In this way, it is possible to resolve the different steps of the breakdown process [5–8], although the use of ex situ optical, electron, or atomic force microscopies is necessary to correlate the electrochemical findings with specific features of the object studied [9–11].

Scanning microelectrochemical techniques provide an alternative for the study of protection methods and degradation reactions with spatial selectivity, especially with regards to the initiation and subsequent propagation of the latter [12–15]. Conceptually, these methods can be considered scanning probe microscopies (SPMs) in which microelectrodes are used as sensing probes in near-field scanning configurations, leading to the monitoring of electrochemical systems that are both chemically and spatially resolved for the surfaces under investigation. In fact, SPM is a branch of microscopy where the pixels of the eventual image are obtained by sequentially implemented local measurements. The main three groups of scanning microelectrochemical techniques are based on the measurement of local potential fields (namely, the scanning reference electrode technique, SRET, and the scanning vibrating electrode technique, SVET) [12], the measurement of local electrochemical impedances (i.e., the localized electrochemical impedance spectroscopy, LEIS [15]), and the electrochemical operation of microelectrodes in the scanning electrochemical microscope (SECM) [14]. Among them, the only technique capable of obtaining chemical resolution and specificity is SECM, since the measuring probe can be configured to specifically monitor a certain chemical species.

Since its introduction in 1989 by Engstrom's [16] and Bard's [17] research groups, SECM has found increasing application in various scientific fields, namely Chemistry, Materials Science, Chemical Engineering, and Corrosion Engineering, among others [18]. As result, the scientific production using SECM has been steadily growing over the years and reporting applications in different research areas. In particular, SECM has also been used for the characterization of organic and inorganic coatings applied on metals, where this technique has found application for microscopic chemical imaging, the measurement of physicochemical constants and coefficients, and as a micromachining tool [19]. A relevant review addressing strategies and operation modes in SECM for the investigation of corrosion processes in metals and their alloys was published by Payne et al. in 2017 [14]. The latter work contained a brief section about the uses of SECM in the investigation of corrosion reactions in coated metals including a summary of different corrosion protection schemes classified by kind of protection and metal.

In 2016, a comprehensive review of experimental parameters in SECM was published by Polcari et al., covering applications in different fields [18], whereas Zoski focused on the use of SECM for surface reactivity characterization, emphasizing novel operation modes, as well as the detection and quantification of metal oxides of the materials used for tip fabrication [20]. Unfortunately, no similar effort has been made in previous reviews on the application of SECM to the study of corrosion processes for covering key experimental parameters of the measurements for data such as the potential applied to the tip, the composition of the measuring solution, tip stability and dimensions, or the eventual effect of redox mediator conversion at the tip on the actual corrosion process under investigation.

The present review addresses the different modes of operation in SECM and their application to the study of coatings and metal-coating systems, with a detailed breakdown of experimental aspects. This work effectively updates and extends our previous report where the operation modes available in SECM for the study of degradation processes in coated metals were described and illustrated [1], also including a review of the experimental parameters involved in the measurements that constitute analytical figures of merit for the comparison and quantitative evaluation of SECM data [21].

2. Experimental Design for SECM Operation

There are several important factors to be considered when designing a SECM experiment. The most important ones are the nature and geometry of the tip that will determine the spatial and chemical resolutions of the measurement, as well as the type of substrate and mediator, and the solvent to be employed. A review of the main experimental design factors related to the application of SECM to the analysis of coating systems is summarized in the next sections. All the information is organized in tables in order to simplify the presentation of the available resources and their easier comparison. We also provide a brief description of the main aspects related to the features described in the next sub-sections. It must be noted that method validation will not be discussed here as it is beyond the reach of this work, but a relevant review on the topic has been recently presented by Izquierdo et al. [21].

2.1. SECM Instrumentation

SECM is a scanning probe microscope (SPM) technique based on electrochemical principles. The movement of a microelectrode (ME) in close proximity to the interface under study allows the application of various experimental procedures that can be classified into amperometric and potentiometric operations depending on either sensing faradaic currents or probe potential values due to concentration distributions resulting from the corrosion process, as sketched in Figure 2. In addition, alternating current signals can be applied to the ME, leading to AC-operation modes.

As sketched in Figure 2A, amperometric SECM operation is performed in a small electrochemical cell constituted by the tip, the counter electrode, the reference electrode, the substrate, and the solution. A bipotentiostat completes the SECM electrochemical setup together with the electrochemical cell, and it can be used to independently control the potential (bias) of the tip and the substrate, although the latter can also be left unbiased at its spontaneous corrosion potential in the environment. Next, a micropositioner is driven by stepper motors or piezoelectric elements to achieve the movement of the tip in the *X*, *Y*, and *Z* directions for exploring the substrate with submicrometric resolution. The experimental setup is completed with the interface, display system, and computer that records the current at the tip (and eventually the substrate when polarized by the bipotentiostat) as a function of tip position or the potential of the corresponding working electrode.

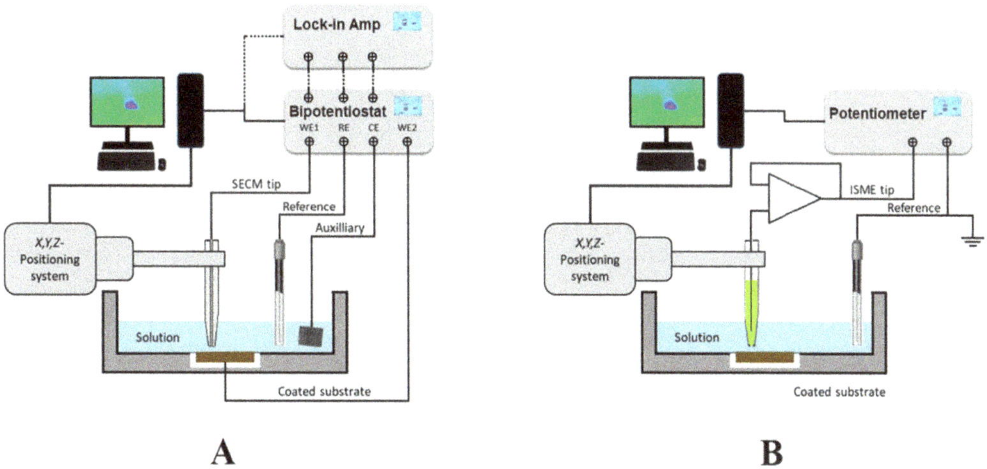

Figure 2. Sketches depicting the SECM set up and electrode connections for (**A**) amperometric and AC operations; and (**B**) potentiometric operation.

AC modes are available by attaching a lock-in amplifier or a frequency response analyser (FRA) to the bipotentiostat, as shown in Figure 2A. 4D AC-SECM mode involves the electrochemical imaging of the AC components of the current signal flowing at the tip (i.e., admittance and phase angle) [22], whereas impedance spectra can be generated at the scanning electrochemical impedance microscope (SEIM) by combining the local current and potential signals [23]. Research applications of these techniques to characterize thin surface layers and coatings on metals are summarized in Table 1.

Table 1. Selected summary of successful applications of AC modes in SECM for the investigation of thin surface layers and coatings on metals.

Technique	Application	Reference
AC-SECM	Visualisation of pin holes on lacquered tinplate	[24,25]
AC-SECM	Imaging of a scratch in polymer-coated galvanized steel	[26]
AC-SECM	Visualization of the adsorption of corrosion inhibitor layers on copper	[27,28]
AC-SECM	Definition of a characteristic threshold frequency during adsorption of corrosion inhibitor layers on copper	[29]
AC-SECM	Water uptake and early coating swelling in coil coated steel	[30]
AC-SECM	Holiday produced in a thin epoxyphenolic varnish applied on tinplate	[31,32]
SEIM	Visualization of the adsorption of corrosion inhibitor layers on copper	[23,33]
SEIM	Alumina layers on aluminium	[34]
AC-SECM	Self-healing performance of smart coatings loaded with corrosion inhibitors	[35,36]

Alternately, potentiometric operation can be performed by measuring local potential signals in a two-electrode cell configuration, as sketched in Figure 2B. In this case, a high input impedance operational amplifier must be introduced between the electrode connections in the electrochemical cell before they are driven to the bipotentiostat or the potentiometer unit employed as electrochemical interface [37]. Ion-selective microelectrodes (ISMEs) are used as SECM probes instead of the active tip surfaces employed in amperometric and AC modes. Although potentiometric SECM, also known as the scanning ion selective electrode technique (SIET) in some publications, has usually been performed using two separate electrodes in the electrochemical cell, as sketched in Figure 2B, it was demonstrated that such an arrangement contributes to big uncertainties in the measurement of local potential values in systems undergoing corrosion reaction, due to the high electrical fields developed in the electrolyte by galvanic pair systems [38]. The use of internal reference electrodes built inside the ion-selective electrode tip should be mandatory in order to overcome this reported limitation [39,40]. Another limitation arises from the rather slow equilibration time required to establish a stationary Donnan potential in the ion-selective membrane of the ISME, effectively limiting the resolution of the chemical images that can be recorded using this operation mode, which is often reduced to a few 2D line scans [41], although new imaging procedures involving the construction of 3D pseudo-maps have recently become available [42].

A summary of the SECM instruments and the analytical figures of merit necessary for adequate description of the experiments and reproducibility are reviewed elsewhere [21].

2.2. Tips Used for Amperometric Operation

Tips employed in amperometric SECM are active microelectrodes (MEs) of a critical dimension below 25 μm for conditioning the mass transport of a redox species from the solution toward the electrode, which are used to characterize coatings and/or thin surface layers applied on metals without requiring any additional modification [43]. In SECM, the most employed tip is built using a platinum wire of different diameters, followed by gold and carbon microwires or fibres, although antimony- and iridium-based tips have been occasionally employed due to the dual amperometric/potentiometric potential of their oxides [44,45].

For the tip fabrication, usually a metal microdisk (e.g., Pt) is sealed into a glass capillary and tapered to a conical shape. Then, it is polished with graded alumina powder (or similar) of different sizes in order to expose a disk-shaped electrode with an active surface for the redox reaction over its surface and is fully characterized by recording a cyclic voltammogram of a known electroactive species (e.g., redox mediator) added to the test solution, as it is exemplified in Figure 3 for the case of the oxidation of ferrocene-methanol on a Pt ME.

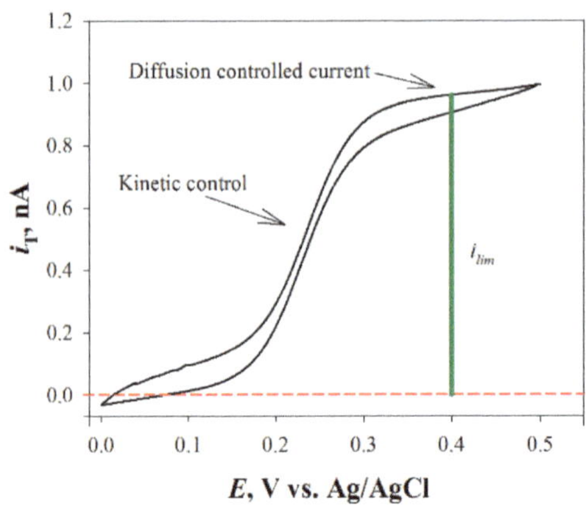

Figure 3. Cyclic voltammetry curve for ferrocene-methanol (i.e., the most employed redox mediator in the literature) on a Pt ME.

The variation of the measured current (i_T) at the surface of the ME with potential makes it possible to obtain relevant data on the active surface and the tip geometry, as well as the value of the stationary diffusion current, $i_{lim} = i_{T,\infty}$, and the operative potential range that is available at the tip. This stationary current is controlled by the electrochemical characteristics of the redox species (namely its diffusion coefficient and concentration) together with the geometrical factors of the tip [46]. The latter are accounted for by the radius of the electroactive surface of the electrode (a, in cm), and a geometric coefficient, β, that depends on the ratio between the diameters of the active disk electrode (2a) and of the insulating shaft built around it (S) that is named the R_g value [46]. This stationary current is measured for the tip placed in the bulk of the solution, and it is effectively observed when the probe is positioned at a distance from the substrate 10 times greater than a.

To compare information from different measurements, SECM data are usually plotted using normalized quantities for the current, I, and for the tip-substrate distance, L. These parameters are obtained as the ratios of the current measured at the tip at some distance from the substrate, d, to the stationary current ($I = i_T/i_{T,\infty}$), and to the tip radius ($L = d/a$), respectively.

Although glass-embedded metallic Pt tips, with a clear predominance of the tips of 10 to 25 μm in diameter, are the most used in this field, in some cases Pt coated with parylene C [47] and a Pt/IrO_x tip [48], both of 25 μm diameter, have been employed. Although tip diameters up to 100 μm have been reported for SECM application [49–51], it must be taken in account that they are too big for applying the analysis tools developed for microelectrode configurations, which are described in Section 3. Other materials used have been Au disks (5 and 25 μm of diameter) [52], and boron-doped diamond (BDD) films on tungsten wires [53].

2.3. Redox Mediators

A redox mediator is an electroactive molecule, atom, or ion that can be reduced or oxidized. Mediators are classified as direct and indirect redox mediators depending on whether the species is already present in the solution (e.g., O_2) or has to be added to the solution to perform the experiment (e.g., ferrocene-methanol, FcMeOH). The latter is needed to monitor either insulating or poorly-conductive substrates, thus reflecting only topographical and/or chemical reactivity changes along a surface, respectively, as sketched in Figure 4. Redox mediators may also be added for precise tip positioning of the tip relative to the substrate by recording the changes in the tip current while approaching the substrate (see Section 3.1), as well as to locate sites of different reactivity on the substrate (cf. Figure 4). Normally, when a redox mediator species is added to the solution, a low concentration (approximately in mM values) is used. However, more strictly, the concentration of the mediator must be chosen by taking into account its reaction rate at the substrate. Besides, it may be necessary to adjust the pH.

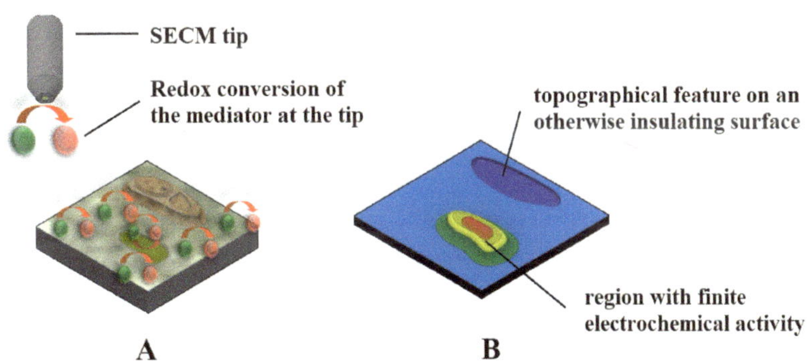

Figure 4. Sketches depicting: (**A**) the redox conversion of an electrochemical mediator at the amperometric tip of SECM over a mostly insulating surface, whereas the redox mediator regeneration solely occurs in a region exhibiting electrochemical reactivity; and (**B**) the current measured at the tip, which contains information on the topography of the sample (i.e., blue color palette) eventually coupled with a region showing a heterogeneous chemical reactivity distribution (red and yellow color palette).

The selection of a certain redox mediator for a given experiment is a critical issue for a successful experiment and depends on several factors, such as:

- The nature of the sample studied;
- The nature of the mediator (chemical stability, redox potential, photostability, toxicity, thermal stability, and solubility in the solution to be tested); and
- The mode of operation to use in SECM.

A broad classification of redox mediator systems for the investigation of electrochemical corrosion processes can be made by considering whether the species is added to the test environment for imaging (ideal redox mediator system) or whether it is a certain chemical species that participates in the corrosion mechanism, although the latter frequently exhibits poorly reversible or even irreversible electron transfer reactions (for instance, the electroreduction of molecular oxygen) and highly variable concentration ranges. Since corrosion reactions on coated metals often expose insulating or poorly conductive layers to the electrolytic phase in which the SECM tip is moved, the use of corrosion-related mediators is limited to systems presenting either defects or cut edges, or low efficiencies of inhibition. In contrast, the addition of redox mediators that exhibit fast, simple, and highly reversible electron transfer reactions at the tip is preferred in the case of non-defective and barrier-type layers and coatings, where the initiation of the corrosion reactions occurs in the buried interface formed by the metal and the surface layer. In this case, the information collected on the degradation process is obtained by observing morphological and topographical changes on the outer surface of the coating or film. An intermediate case occurs during the formation of inhibiting layers on metals by adsorption, because there is a gradual transition from an electrochemically active surface to a (quasi) insulating surface, making it possible to follow the time course of the charge transfer reaction at the surface [54]. In this case, the main care must relate to the selection of a mediator having a redox potential close to the corrosion potential of the substrate, in order to minimize the effects of polarization of the substrate by the redox couple.

Although at least 133 molecules have been used or investigated for amperometric operation in SECM since its inception [18], Table 2 shows that only a few of them have been used for SECM characterization of thin surface layers and coatings on metals.

Table 2. Redox mediators used in the amperometric SECM characterization of thin surface layers and coatings on metals.

Mediator	Abbreviation	Redox Reaction	Redox Potential (V vs. NHE)	Reference
Azobenzene	AB	$AB + e^- \rightarrow AB^{\bullet -}$	+1.378	[18]
Benzoquinone/hydroquinone	BQ/HQ	$BQ + 2H^+ + 2e^- \rightarrow HQ$	−0.278	[55]
Decamethylferrocene	DcMeFc	$[DcMeFc]^+ + e^- \rightarrow DcMeFc$	+0.261	[56]
Dimethylamino-methylferrocene	DMAMFc	$[DMAMFc]^+ + e^- \rightarrow DMAMFc$	+0.551	[57]
Ferrocene	Fc	$Fc^+ + e^- \rightarrow Fc$	+0.665	[58]
Ferrocenemethanol	FcMeOH	$[FcMeOH]^+ + e^- \rightarrow FcMeOH$	+0.500	[59]
Hexaammineruthenium (III)	$[Ru(NH_3)_6]^{3+}$	$[Ru(NH_3)_6]^{3+} + e^- \rightarrow [Ru(NH_3)_6]^{2+}$	−0.059	[60]
Hexacyanoferrate (III)	$[Fe(CN)_6]^{3-}$	$[Fe(CN)_6]^{3-} + e^- \rightarrow [Fe(CN)_6]^{4-}$	+0.491	[61]
Hydrogen	H_2	$2H^+ + 2e^- \rightarrow H_2$	0.000	[18]
Hydrogen peroxide	H_2O_2	$O_2 + 2H^+ + 2e^- \rightarrow H_2O_2$	+0.670	[62]
Iodide	I^-	$I_3^- + 2e^- \rightarrow 3I^-$	+0.963	[63]
Iodine	I_2	$I_2 + 2e^- \rightarrow 2I^-$	+0.532	[64]
Iridium chloride	$IrCl_6$	$[IrCl_6]^{2-} + e^- \rightarrow [IrCl_6]^{3-}$	+0.870	[65]

Table 2. *Cont.*

Mediator	Abbreviation	Redox Reaction	Redox Potential (V vs. NHE)	Reference
Iron	Fe	$Fe^{3+} + e^- \rightarrow Fe^{2+}$	+0.772	[66]
Methylviologen	MV	$MV^{2+} + e^- \rightarrow MV^+$	−0.446	[67]
4-nitrobenzonitrile	4NB	$4NB + e^- \rightarrow 4NB^{\bullet-}$	+0.659	[68]
Oxygen	O_2	$O_2 + 2H_2O + 4e^- \rightarrow 4OH^-$	+0.401	[62]
Oxygen	O_2	$O_2 + e^- \rightarrow O_2{}^-$	−0.498	[69]
Tetramethyl-p-phenylenediamine	TMPD	$TMPD^+ + e^- \rightarrow TMPD$	+0.258	[70]
Tetracyanoquinodimethane	TCNQ	$TCNQ + e^- \rightarrow TCNQ^-$	+0.322	[18]
Tetrathiafulvalene	TTF	$TTF^{2+} + e^- \rightarrow TTF^+$ $TTF^+ + e^- \rightarrow TTF$	+0.593 +0.943	[64]

2.4. *Tips Used for Potentiometric Operation*

Passive tips (ISME) are employed in potentiometric SECM. Since ion activities are detected using ISME tips, without being consumed during the measurement, no interaction must occur with the sample surface. Although chemical selectivity is thus envisaged compared to amperometric operation, the selectivity of the probe is not always sufficiently high and it is necessary to be investigated in regards to other ions present in the system [71]. Additionally, longer acquisition times are required for SECM measurement due to longer response times of the probes. When scanning rates similar to those typical for amperometric SECM operation are employed, the recorded images may exhibit significant aberration effects. Nevertheless, significant improvement has been achieved by combining dedicated scanning routines with mathematical deconvolution procedures [72,73].

Potentiometric probes used in SECM can be classified into two broad categories, namely metal-based microsensors [74] and reference microelectrodes [75]. The first class of potentiometric microsensors take advantage of the passive properties of certain metal oxides that are primarily sensitive mainly to pH changes in the environment, such as antimony and iridium. Although the narrow potential range of stability of its metallic state in electrolyte solutions precludes its use as an electrode material with voltammetric techniques, it encompasses the potential range of O_2 electroreduction and therefore can be used in SECM for precise probe positioning [76,77], which is a typical limitation of most passive potentiometric probes. In this way, the *z*-approach curves are recorded with the metal in the active state (i.e., operating as a conventional amperometric SECM tip with redox mediators that are converted within the potential stability range of the elemental state of the tip metal), and then oxidized to produce the pH-sensitive metal oxide layer [78]. In other cases, positioning is achieved with dual microelectrodes with the feedback mode by adding a redox mediator to the electrolyte [45,79], a procedure that will be described in Section 3.1. Unfortunately, only a very small number of ion species can be detected using this type of microsensors (namely, H^+, Ag^+, and Cl^-), and indeed they have only been used as pH microscopy when characterizing thin surface layers and coatings on metals, as shown in Table 3.

Table 3. Metallic microsensors used as potentiometric probes in SECM characterization of thin surface layers and coatings on metals.

Ion	Metal	Application	Reference
H^+	Antimony-antimony oxide	Corrosion reactions at cut edges of galvanized steel and polymer coated galvanized steel	[42,77,80,81]

Table 3. *Cont.*

Ion	Metal	Application	Reference
H^+	Antimony-antimony oxide	Corrosion inhibition efficiency of 2-mercaptobenzothiazole on copper	[82]
H^+	Antimony-antimony oxide	Corrosion inhibition efficiency of benzotriazole for the galvanic coupling of copper and iron	[83]
H^+	Iridium-iridium oxide	Corrosion reactions at scratched alkyd-melamine coating applied on 16 MnCrS5 carbon steel	[84]
H^+	Platinum-iridium oxide	Corrosion reactions on 316 L stainless steel surface	[79,85]

The second type of passive potentiometric probes corresponds to ion-selective microelectrodes (ISMEs), which consist of a selective transducer (usually a membrane) that transfers the ion activity of a certain species occurring in the electrolyte phase to an electrical potential. The sensing membrane is a multicomponent solution (herein named cocktail) containing the ionophore, the polymeric matrix, the lipophilic ion exchanger, and the lipophilic salt. The ionophore is the component that selectively forms a complex with the primary ion to be monitored, whereas the polymeric matrix accounts for the mechanical stability of the system. Since membranes must be immiscible with water, lipophilic components are employed.

Although a detailed review on the use of potentiometric probes as SECM sensors can be found elsewhere [86], Table 4 lists the ISME employed to monitor thin surface layers and coatings on metals.

Table 4. Ion-selective microelectrodes (ISME) used as potentiometric probes in SECM characterization of thin surface layers and coatings on metals.

Ion	Ionophore	Ion-Selective Cocktail	Application	Reference
H^+	Hydrogen ionophore I	Cocktail B	Corrosion and self-healing functions at cut-edges of galvanized steel	[87,88]
H^+	Hydrogen ionophore I	Cocktail B	Imaging of microdefects in sol-gel film coatings deposited on AZ31 and AZ31B magnesium alloys	[89,90]
H^+	Hydrogen ionophore I	Cocktail B	Corrosion protection of inhibitor loaded composite coatings on AZ31 magnesium alloy	[91]
H^+	Hydrogen ionophore II	Potassium tetrakis (4-chlorophenyl)borate and membrane solvent 2-nitrophenyloctyl ether	Inhibitor-doped hydroxyapatite (HA) microparticles applied over aluminum alloy (AA2024)	[92]
H^+	Hydrogen ionophore I	Cocktail B	Particulate 6092-T6 Al metal matrix composites reinforced with 20 vol.% of B_4C, SiC, and Al_2O_3	[93,94]
H^+	Hydrogen ionophore I	Cocktail B	Galvanic corrosion and localized degradation of aluminium-matrix composites reinforced with silicon particulates	[95]
H^+	Hydrogen ionophore II	Potassium tetrakis(4-chlorophenyl)borate and membrane solvent 2-nitrophenyloctyl ether	Smart coatings applied to galvanized steel	[96]
H^+	Hydrogen ionophore II	Cocktail A	Steel samples with Al-Zn-Mg coatings	[97]

Table 4. *Cont.*

Ion	Ionophore	Ion-Selective Cocktail	Application	Reference
H^+	Hydrogen ionophore I	Cocktail B	Cut edge consisting of a zinc anode and a split iron cathode	[98]
Mg^{2+}	Magnesium ionophore II	Cocktail B	AZ31 and AZ31B magnesium alloys coated with a thin sol–gel film	[89,90]
Mg^{2+}	Magnesium ionophore II	Cocktail B	Corrosion protection of inhibitor loaded composite coatings on AZ31 magnesium alloy	[91]
Mg^{2+}	Bis-N,N-dicyclohexyl-malonamide	Tetrahydrofurane, poly(vinyl chloride), potassium tetrakis(4-chlorophenyl)-borate, and 2-nitrophenyl octyl ether	Galvanic corrosion of Mg coupled to Fe	[40,99,100]
Mg^{2+}	N;N′,N′′-tris[3-(heptylmethylamino)-3-oxopropionyl]-8,8′-iminodioctylamine	Potassium tetrakis(4-chlorophenyl) borate, o-nitrophenyl-n-octylether, poly(vinylchloride), and cyclohexanone	Galvanic corrosion of Mg coupled to Fe	[101]
Mg^{2+}	Magnesium ionophore II	Cocktail A	Galvanic corrosion of Mg coupled to Fe	[102]
Zn^{2+}	Zinc ionophore I	Tetrahydrofurane, 2-nitrophenyl octyl ether, poly(vinyl chloride), and potassium tetrakis(4-chlorophenyl)borate	Corrosion reactions at cut edges of galvanized steel and polymer coated galvanized steel	[39,42,76]
Zn^{2+}	Zinc ionophore I	Tetra-n-butyl thiuram disulfide, sodium-tetrakis[3,5-bis(trifluoro-methyl)phenyl]borate, tetrakis(4-chlorophenyl)borate, and tetradodecylammonium, dissolved in 2-nitrophenyloctyl ether	Painted electrogalvanized steel with two artificial defects	[103]
Cl^-	Chloride ionophores I and II	Solvents: 2-nitrophenyl octyl ether, 2-nitrophenyl pentyl ether, 2-nitrophenyl phenyl ether, 1,2-dimethyl-3-nitrobenzene, 2-fluorophenyl 2-nitrophenyl ether, benzyl 2-nitrophenyl ether. Ion exchangers: potassium tetrakis(4-chlorophenyl) borate, and tridodecylmethylammonium chloride	Cut-edge of metallic coated steel	[104]
Na^+	Sodium ionophores II, VI, VIII and X	Solvents: 2-nitrophenyl octyl ether, 2-nitrophenyl pentyl ether, 2-nitrophenyl phenyl ether, 1,2-dimethyl-3-nitrobenzene, 2-fluorophenyl 2-nitrophenyl ether, benzyl 2-nitrophenyl ether. Ion exchangers: potassium tetrakis(4-chlorophenyl) borate, and tridodecylmethylammonium chloride (TDDMACl)	Cut-edge of metallic coated steel	[104]

Chloride ionophore I (24897): meso-Tetraphenylporphyrin manganese(III)-chloride complex [$C_{44}H_{28}ClMnN_4$]. Chloride ionophore II (24901): 4,5-Dimethyl-3,6-dioctyloxy-o-phenylene-bis(mercurytrifluoroacetate) [$C_{28}H_{40}F_6Hg_2O_6$], ETH 9009. Cocktail A: 2-Nitrophenyl octyl ether, 89.3 wt.% (73732) + Sodium tetraphenylborate, 0.7 wt.% (72018). Cocktail B: 2-Nitrophenyl octyl ether, 89.3 wt.% (73732) + Potassium tetrakis(4-chlorophenyl)borate, 0.7 wt.% (60591). Hydrogen ionophore I (95292): Tridodecylamine [$CH_3(CH_2)_{11}]_3N$. Hydrogen ionophore II (95295): 4-Nonadecylpyridine ($C_{24}H_{43}N$), ETH 1907, Proton ionophore II. Magnesium ionophore II (63083): N,N′′-Octamethylene-bis(N′-heptyl-N′-methyl-methylmalonamide [$C_{32}H_{62}N_4O_4$], ETH 5214. Sodium ionophore II (71733): N,N′-Dibenzyl-N,N′-diphenyl-1,2-phenylenedioxydiacetamide [$C_{36}H_{32}N_2O_4$], ETH 157. Sodium ionophore VI (71739): Bis[(12-crown-4)methyl] dodecylmethylmalonate, Dodecylmethylmalonic acid bis[(12-crown-4)methyl ester] [$C_{34}H_{62}O_{12}$]. Sodium ionophore VIII (73929): Bis[(12-crown-4)methyl] 2,2-didodecylmalonate [$C_{45}H_{84}O_{12}$]. Sodium ionophore X (71747): 4-tert-Butylcalix[4]arene-tetraacetic acid tetraethyl ester [$C_{60}H_{80}O_{12}$]. Zinc ionophore I (96491): Tetrabutylthiuram disulfide [$C_{18}H_{36}N_2S_4$].

3. Operation Modes

As described above, the foundation of amperometric operation is the change in the measured current (i_T) at the surface of a biased microelectrode occurring when it is moved near the surface of a substrate immersed in an electrolyte solution containing a redox mediator. Different operation modes can be identified depending on the origin and function of the redox mediator in the electrochemical system formed by the tip and the investigated substrate.

3.1. Feedback Modes

The feedback mode was one of the first operation modes employed in SECM [105], and it is one of the most frequently employed modes due to its versatility. In this mode, the tip current (i_T) due to the redox conversion of a redox mediator is monitored, and its magnitude varies with the tip/substrate distance (d), the chemical nature of the mediator as well as the composition and conductivity of the electrolytic solution. A different potential value must be applied to the tip for each redox mediator. In the presence of a non-conductive substrate, the diffusion of the mediator is hindered and eventually blocked in the proximity of the substrate. That is, the faradaic current measured at the tip, i_T, gradually decreases while performing an approach of the tip to the substrate because the diffusion of the mediator towards the active area of the tip is hindered by the proximity of the substrate ($i_T < i_{T,\infty}$), and this behaviour is named negative feedback (see Figure 5A). Since the underlying metal is not in direct contact with the electrolyte medium in the case of non-defective insulating coatings, the use of a redox mediator and its eventual development of a redox potential in the system produces no significant effect on the investigated system. Although the information provided by the technique has only spatial resolution (i.e., topography and morphology), this mode has found application for the investigation of transport phenomena through defect-less barrier organic coatings applied on metallic substrates leading to mechanistic information on water uptake [106–108] and lixiviation processes [109], as well as the detection of the early stages of coating blistering and delamination induced by ionic species such as chloride [110–113].

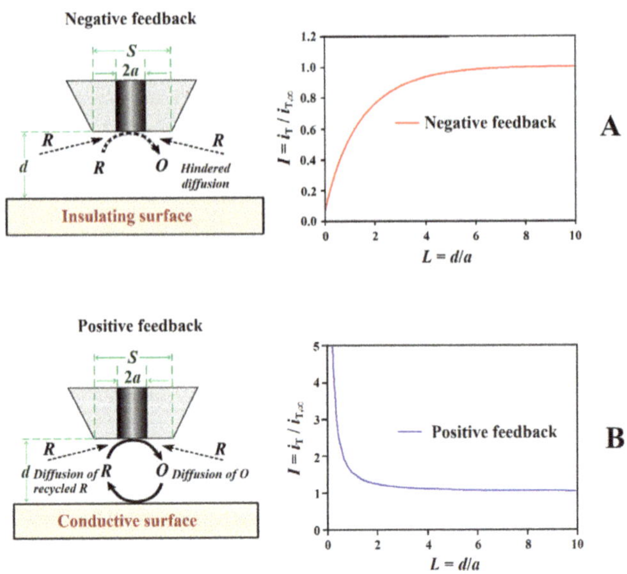

Figure 5. Schemes and shapes of Z-approach curves in the feedback mode of SECM [114]. Types of feedback: (**A**) negative and (**B**) positive.

Conversely, if the surface of the sample is conductive, the mediator can be regenerated on it, and an increase of i_T can be observed ($i_T > i_{T,\infty}$) for smaller tip-substrate distances originating a positive feedback behaviour (cf. Figure 5B). The study of electrically insulating or conductive surfaces is possible thanks to the appearance of negative or positive feedback effects, obtaining images of the studied surface that reflect the occurrence of defects in insulating coatings, including both inorganic and organic matrices, to be investigated [41,115–117]. Depending on the size of the tip, the measurement of i_T can thus provide information about sample topography and its electrical and chemical properties, allowing for the occurrence of defects ranging from pinholes to holidays and scratches, to be detected, as well as to monitor their evolution [115–117]. In brief, insulating regions produce changes in the current measured at the tip due to topographic changes that modify the transport regime of the redox mediator from the electrolyte bulk towards the tip, frequently interfering with the signal while scanning the substrate in close proximity. Conversely, regions in the substrate that are conductive and capable of regenerating the redox mediator produce an increase in current measured at the tip. It should be noted that, since the current response in feedback mode is highly dependent on tip-to-substrate distance, it is preferable to use as small a distance as possible (without crashing the tip) to increase sensitivity.

In the case of the feedback operation mode, kinetic information can also be extracted from the experimental approach curves after taking in account the geometric factors of the tip [118], which is of direct application for the determination of the rate constants associated with the formation of corrosion inhibitor layers on metals and their ageing by thickening or greater compactness [119–123]. Additionally, the method has also been employed to gain information on the adsorption of the inhibitor molecules on the metal [54].

A total of 18 different mediators have been employed until now in feedback operation, as listed in Table 5. The most used is ferrocene-methanol (FcMeOH) (ca. 36% of the cases) followed by hexacyanoferrate (III), $[Fe(CN)_6]^{3+}$, (with almost 17% of the cases). This fact is all the more striking as a wide variety of substrates are involved in the studies, and their characteristic potential values in the test environment vary considerably from highly reactive metals such as magnesium alloys to more noble materials such as copper. That is, it is often not taken into account that the coexistence of the two forms of a redox mediator confers a potential on the sample that effectively acts as if an external bias were applied, so that the same mediator should not be the preferred choice for widely dissimilar metals.

Table 5. Redox mediators used in SECM feedback mode for the characterization of thin surface layers and coatings on metals.

Mediator	Electrolyte Solution	E_{tip}	Substrate	Coating	Tip (φ, μm)	Reference
AB	1 mM AB + 0.1 M Bu_4NBF_4 in 0.1 M ACN	−1.6 V vs. SCE	Glass	Poly(3,4-ethylenedioxythiophene) (PEDOT)	Pt (10)	[124]
H^+	1 mM BQ + 0.1 M Bu_4NBF_4 in 0.1 M ACN	−0.5 V vs. SCE	Glass	PEDOT	Pt (10)	[124]
H^+	0–50 mM p-BQ M in PC	−1.0 V vs. NHE	Glass	AgCl-coated sample	Pt (10)	[125]
H^+	ethyleneglycol + 10–50 mM p-BQ + 20 mM KI		Teflon			
DcMeFc	0.5 mM DcMeFc + 5 mM BATB in DCE	-	Pt bands over glass	Parylene C	Pt_{pC} (25)	[47]

Table 5. *Cont.*

Mediator	Electrolyte Solution	E_{tip}	Substrate	Coating	Tip (φ, μm)	Reference
DMAMFc	1.25 mM $DMAMFc^+$ + 10 mM H_2SO_4	+0.40 V vs. Ag/AgCl	AA2024-T3 aluminum alloy	Poly(aniline) + poly-(methylmethacrylate) (PANI-PMMA) blend	Pt (10)	[126]
	1.25 mM DMAMFc + 10 mM BBS	+0.80 V vs. Ag/AgCl	AA2024-T3 aluminum alloy	Zr(IV)–alkyl-phosphonate	Pt (10)	[127]
		+0.16 V vs. SCE		Zr(IV)–aryl-phosphonate		
Fe	0.01 M Fe^{3+} + 0.5 M H_2SO_4	−0.1 V vs. Ag/AgCl	-	Polyester-polypyrrole-graphene oxide (PPy/GO)	Pt (100)	[51]
Fc	1 mM of Fc + 0.1 M Bu_4NBF_4 in ACN	+0.4 V vs. SCE	Glass	PEDOT	Pt (10)	[124]
	0.1 M Bu_4NBF_4 + 0.1 M in ACN	+0.41 V vs. Ag/AgCl	Glass	Zn-porphyrin layers on indium tin oxide (ITO) electrode and ITO modified with poly-1	Pt (10)	[128]
FcMeOH	0.5 mM FcMeO + 0.1 M KCl	+0.5 V vs. Ag/AgCl/KCl sat.	Carbon steel (CS)	Two-component polyurethane	Pt (10)	[66]
FcMeOH	0.5 mM FcMeOH + 0.1 M NaCl	+0.8 V vs. Ag/AgCl	AA2024 aluminum alloy	1% γ-aminopropyltrimethoxy silane (γ-APS)-doped and 2.5% bis-1,2-[triethoxysilyl]ethane (BTSE)-doped epoxy coating	Pt (25)	[129]
FcMeOH	0.9 mM FcMeOH + 5 wt.% NaCl	-	Q235 mild steel (MS)	Enamel coating	-	[130]
FcMeOH	0.5 mM FcMeOH + 3.5 wt.% NaCl	+0.5 V vs. Ag/AgCl/KCl sat.	High strength steel (SAPH440)	Enamel coating	-	[131]
FcMeOH	5 mM FcMeOH + 0.05 M NaCl	+0.6 V vs. Ag/AgCl/KCl sat.	AA2024-T3 aluminum alloy	Epoxy coating and epoxy coating containing silyl-ester doped capsules	Pt (5)	[132]
FcMeOH	5 mM FcMeOH + 0.5 mM NaCl	+0.5 V vs. Ag/AgCl/KCl sat.	Coil coated steel (CCS)	Polyester paint	Pt (10)	[106]
FcMeOH	0.1 M TBATFB in PC or ACN	+0.4 V vs. Ag/AgCl	Gold-coated silicon and p-Si	Oligothiophenes	Pt (25)	[133]
FcMeOH	1 mM FcMeOH + 0.1 M KNO_3	+0.35 V vs. Ag/AgCl	2024-T3 aluminum alloy	Non-chromated primer on anodized Al; waterborne primer on alodine pretreated Al; chromated primer on alodine pretreated Al	Pt (25)	[134]
FcMeOH	0.9 mM FcMeOH + 0.1 M KNO_3	+0.4 V vs. Ag/AgCl	Silicon	Cu_xS substrate + SiO_2/Si	Pt (10)	[135]

Table 5. *Cont.*

Mediator	Electrolyte Solution	E_{tip}	Substrate	Coating	Tip (φ, µm)	Reference
FcMeOH	10 mM FcMeOH + 0.1 M TBAPF6 in DMF	-	Pt	PEDOT	Pt (30)	[136]
FcMeOH	2.2 mM FcMeOH + 0.1 M KNO_3	+0.2 V vs. Ag-QRE	Pt bands over glass	Parylene C	Pt_{pC} (25)	[47]
FcMeOH	0.5 mM FcMeOH + 0.05 M NaCl	+0.4 V vs. Pt-wire	Nickel foil	Plasticized Polyvinyl chloride (PVC)	Pt (10)	[109]
FcMeOH	1 mM FcMeOH + 0.1 M KNO_3	+0.4 V vs. Ag/AgCl	Cu	Monolayer of $C_{12}H_{25}$–X (X = –SH, –S–S–, –SeH and –Se–Se–)	Pt (10)	[137]
FcMeOH	1 mM FcMeOH + 0.1 M KNO_3 or K_2SO_4	+0.4 V vs. Ag/AgCl	Copper-based quaternary bronze (UNS C83600)	Cu Patina	Pt (25)	[138]
FcMeOH	1 mM FcMeOH + 0.1 M KCl	+0.4 V vs. Ag/AgCl	-	Polytetrafluorethylene (PTFE)	BDD (6 to 23)	[53]
FcMeOH	1 mM FcMeOH + 1 mM Na_2SO_4	+0.50 V vs. Ag/AgCl/KCl (3 M)	Cu	2-Mercaptobenzimidazole (2-MBI)	Pt (10)	[123]
FcMeOH	Ringer's physiological solution	+0.47 V vs. SCE	Ti-6Al-4V and Ti-21Nb-15Ta-6Zr alloys	HA–ZrO_2	Pt (12.5)	[116]
FcMeOH	0.67 mM FcMeOH + 0.067 M Na_2SO_4 + 0.33 mM BTAH	+0.50V vs. Ag/AgCl/KCl (3 M)	Cu	Benzotriazole (BTAH)	Pt (25)	[122]
FcMeOH	1 mM FcMeOH + 0.2 M KCl	+0.4 V vs. Ag/AgCl	AA2024-T3 aluminum alloy	Epoxy resin + vanadate- and tungstate-doped PPy/Al flake composite pigments	Pt (10)	[139]
FcMeOH	1 mM FcMeOH + 0.1 M K_2SO_4	+0.6 V vs. Ag/AgCl	Inconel 625	Thin coatings of the alloy on MS using High velocity oxy-fuel (HVOF)	Pt (4)	[140]
FcMeOH	5 mM FcMeOH + 0.01 M NaCl	+0.5 V vs. Ag/AgCl/KCl sat.	2024 aluminium alloy	Acrylic coat (undoped coating system) and with mesoporous pretreatment	Pt (10)	[141]
FcMeOH	0.5 mM FcMeOH + 0.1 M KCl	+0.5 V vs. Ag/AgCl, KCl sat.	MS	Polyester	Pt (10)	[13]
FcMeOH	0.5 mM FcMeOH + 0.1 M KCl	+0.50 V vs. Ag/AgCl, KCl sat.	MS	Polyester	Pt (10)	[142]
FcMeOH	0.5 mM FcMeOH + 0.1 M KCl and 0.5 mM FcMeOH + 0.1 M K_2SO_4	+0.5 V vs. Ag/AgCl	MS	Polyurethane	Pt (10)	[111]

Table 5. *Cont.*

Mediator	Electrolyte Solution	E_{tip}	Substrate	Coating	Tip (φ, μm)	Reference
FcMeOH	1 mM FcMeOH + 0.1 M Na_2SO_4	-	MS	Inconel 625 formed using a HVOF	Pt (10)	[143]
FcMeOH	1 mM FcMeOH + 0.1 M NaCl	-	Q235 MS	Graphene oxide–mesoporous silicon dioxide layer–nanosphere structure loaded with tannic acid (GSLNTA)	Pt (25)	[36]
FcMeOH	1 mM FcMeOH + 0.1 M KCl	-	5083 aluminum alloy	Hexamethylene diisocyanate trimer (HDIt) microcapsules into epoxy	Pt (25)	[144]
FcMeOH	0.9 mM FcMeOH + 0.5 M NaCl	+0.5 V Ag/AgCl	CS	Organosol	Pt (10)	[145]
FcMeOH	0.5 mM FcMeOH + 0.1 M or 0.62 M NaCl	+0.45 V vs. Ag/AgCl/KCl (3 M)	AZ91D magnesium alloy	Surface layer formed by Micro-arc oxidation (MAO)	Pt (10)	[146]
FcMeOH	1 mM FcMeOH + 100 mM Na_2SO_4 and 0.667 mM FcMeOH + 67 mM Na_2SO_4 + 0.333 mM BTAH	+0.45 V vs. SCE	Cu	BTAH	Pt (25)	[119]
FcMeOH	2 mM FcMeOH + 0.2 M Na_2SO_4 with 0.2% (*v*/*v*) ethanol	-	Cu	Poly(3-ethoxy-thiophene) (PEOT) and poly(ethylenedioxy-thiophene) (PEDT)	Pt (10 and 25)	[147]
FcMeOH	1 mM FcMeOH + 0.1 M NaCl	+0.4 V vs. SCE	Cu	Self-assembled monolayers (SAMs) formed by HL	Pt (25)	[148]
FcMeOH	0.5 mM FcMeOH + 0.1 M KCl or 0.1 M K_2SO_4	+0.5 V vs. Ag/AgCl/KCl sat.	CCS	Polyester (PES)	Pt (10)	[107]
FcMeOH	0.5 mM FcMeOH + 0.1 M PBS	0.6 V vs. Ag/AgCl	MS	Electrodeposited silica	Pt (10)	[149]
$[Fe(CN)_6]^{3-}$	1 mM $K_4[Fe(CN)_6]$	+0.4 V vs. Ag/AgCl/KCl sat.	CS	Two-component polyurethane	Pt (10)	[66]
$[Fe(CN)_6]^{3-}$	0.1 M KCl	+0.45 V vs. Ag/AgCl/KCl sat.	Pt and SS	Amorphous alumina thin film grown by Metal organic chemical vapour deposition (MOCVD) process	Pt (25)	[115]
$[Fe(CN)_6]^{3-}$	0.5 mM Ferrocyanide + 44 mM PBS	+0.5 V vs. Ag/AgCl/KCl (3 M)	Gold disk electrode	Cytochrome C	Pt (25)	[150]
$[Fe(CN)_6]^{3-}$	10 mM $K_3Fe(CN)_6$ + 0.1 M Na_2S0_4	−0.4 V vs. SCE	Pt	Nafion film containing Os(pbsy)	Pt (10.1)	[151]

Table 5. *Cont.*

Mediator	Electrolyte Solution	E_{tip}	Substrate	Coating	Tip (φ, µm)	Reference
$[Fe(CN)_6]^{3-}$	1 mM $K_4[Fe(CN)_6]$ + 0.1 M NaCl	+0.40 V vs. Ag/AgCl, KCl sat.	GS	Two-component epoxy primer containing zinc phosphate	Pt (10)	[152]
$[Fe(CN)_6]^{3-}$	50 mM $K_3[Fe(CN)_6]$ + 1 M Na_2SO_4	−0.6 V vs. Ag/AgCl	Pt/Glass	PVC over Pt sheet; poly-terthiophene on glass	Au (5 and 25)	[52]
$[Fe(CN)_6]^{3-}$	10 mM $K_4[Fe(CN)]_6$ + 3 wt.% NaCl 3%	-	Steel	Epoxy resins (Diglycidylether of Bisphenol A (DGEBA) + Methylpentanedi-amine (DAMP))	-	[153]
$[Fe(CN)_6]^{3-}$	2 mM $Fe(CN)_6^{3-}/Fe(CN)_6^{4-}$ + 0.1 M KCl	+0.5 V vs. Ag/AgCl/KCl sat.	Au	Thioglycolic acid (TGA) + Quercetin (Q)	Pt (10)	[154]
$[Fe(CN)_6]^{3-}$	5 mM $K_4[Fe(CN)]_6$ + 0.1 M KBr	+0.5 V vs. Ag/AgCl	Pt	Polytetrafluorethylene (PTFE) plaques	Pt (50)	[49]
$[Fe(CN)_6]^{3-}$	4 mM $K_4[Fe(CN)]_6$ + 0.1 M KNO_3	-	Pt	Pt/400-nm thick layer of a-Si:H Langmuir-Blodgett films of iron oxides nanoparticles	Pt (5)	[155]
$[Fe(CN)_6]^{3-}$	5 mM $Fe(CN)_6^{3-}$ + 0.1 M KCl	-	-	Vinylic monomers	Pt (25 and 100)	[50]
$[Fe(CN)_6]^{3-}$	0.1 M KCl	+0.5 V vs. Ag/AgCl	Glassy carbon substrate electrodes (GCEs)	Bismuth film	Pt (25)	[156]
$[Fe(CN)_6]^{3-}$	5 mM $K_4[Fe(CN)]_6 \cdot 3H_2O$ + 0.1 M KCl	+0.5 V vs. Ag/AgCl	Low CS Q-Panel S	Epoxy resin with Tetraethoxysilane (TEOS) and epoxy resin filled GO	Pt (10)	[157]
$[Fe(CN)_6]^{3-}$	1 mM $K_4[Fe(CN)]_6$ + 0.1 M KPF6	+0.5 V vs. Ag/AgCl	Gold	Azido-terminated self-assembled monolayers	Pt (10)	[158]
$[Fe(CN)_6]^{3-}$	5 mM $K_3[Fe(CN)]_6$ + 100 mM KCl	−0.25 V vs. Ag/AgCl/KCl sat.	Steel	Physical vapour deposition (PVD) TiN coatings	Pt (15)	[159]
$[Fe(CN)_6]^{3-}$	10 mM $K_4[Fe(CN)]_6$ + 3 wt.% NaCl	-	Steel	Epoxy resin with and without TiO_2	Pt (10)	[112]
$[Fe(CN)_6]^{3-}$	10 mM $Ru(NH_3)_6Cl_3$, $K_3Fe(CN)_6$ and $K_4Fe(CN)_6$ + 0.1 M KCl	+0.4 V vs. Ag/AgCl/KCl (3 M)	-	Polyester coated with reduced GO	Pt (100)	[160]
$[Fe(CN)_6]^{3-}$	0.5 mM $K_4[Fe(CN)_6]$ + 3.5% NaCl	+0.40 V vs. Ag/AgCl/KCl (1 M)	CS	Epoxy with zinc phosphate pigment	Pt (10)	[161]
H_2	0.5 mM $HClO_4$ and 0.3 mM $HClO_4$ + 0.1 M $LiClO_4$	−0.65 V vs. Ag/AgCl	Stainless steel (SS)	Pt-TiO_2 prepared by MOCVD	Pt (25)	[162]

Table 5. *Cont.*

Mediator	Electrolyte Solution	E_{tip}	Substrate	Coating	Tip (φ, μm)	Reference
H_2	0.1 M HCl	0.0	Q235 CS	Epoxy coating sample containing zeolitic imidazole framework (ZIF-7)	-	[163]
I^-, I_2	NaCl + KI (no concentrations specified)	-	X80 pipeline steel	Epoxy resin E51 and polyether amine D230	Pt (10)	[164]
$IrCl_6$	1 mM $IrCl_6^{2-}$ + 0.1 M KCl	-	-	PTFE	BDD (6 to 23)	[53]
MV	5 mM MV + 0.1 M KCl	−0.9 V vs. Ag/AgCl	Glass	Composite silica glass containing copper salts	Pt (10)	[165]
O_2	0.1 M KCl	−0.70 V vs. Ag/AgCl/KCl sat.	CS	Two-component polyurethane film	Pt (10)	[66,166]
O_2	2.2 mM FcMeOH + 0.1 M KNO_3	−0.8 V vs. Ag-QRE	Pt bands over glass	Parylene C	Pt_{pC} (25)	[47]
O_2	PBS pH 7.1 + 0.1 M NaCl + 0.01 M NaH_2PO_4	−0.75 V vs. Ag/AgCl, KCl sat.	FTO	Octadeciltrichlorosilane (OTS) based SAMs	Pt (20)	[167]
O_2	0.1 M KF	−0.70 V vs. Ag/AgCl	magnesium mechanically reinforced by powder metallurgy Mg(PM)	Fluorine conversion coatings	Pt (10)	[168]
O_2	3.5 wt.% NaCl	−0.70 V vs. Ag/AgCl/KCl sat.	CS	Epoxy-ZrO_2	Pt (10)	[169]
O_2	0.1 M KCl	−0.70 V vs. Ag/AgCl/KCl sat.	MS	Polyurethane	Pt (10)	[170]
O_2	0.5 mM FcMeOH + 0.1 M KCl	+0.50 V vs. Ag/AgCl/KCl sat.	MS	Polyester	Pt (10)	[142]
O_2	0.1 M KCl + 0.5 mM FcMeOH and 0.1 M K_2SO_4 + 0.5 mM FcMeOH	−0.6 V vs. Ag/AgCl	MS	Polyurethane	Pt (10)	[111]
O_2	3.5 wt.% NaCl	−0.7 V vs. SCE	CS	Styrene-acrylic + terpolymer	-	[171]
O_2	3.5 wt.% NaCl	−0.7 V vs. Ag/AgCl/KCl sat.	AA7075	Berberine	Pt (10)	[172]
4 NB	0.1 M NBu_4PF_6 in ACN	−0.60 V vs. Ag/AgCl	Glass	Zn-porphyrin layers on ITO electrode and ITO modified with the poly-1	Pt (10)	[128]
TMPD	0.1 M $(TBA)BF_4$ + 0.76 mM TMPD in ACN	-	Au	Fullerene	Pt (25)	[173]
TCNQ	1 mM TCNQ + 0.1 M NBu_4BF_6 in ACN	+0.1 V vs. SCE	Glass	PEDOT	Pt (10)	[124]

Table 5. *Cont.*

Mediator	Electrolyte Solution	E_{tip}	Substrate	Coating	Tip (φ, μm)	Reference
TCNQ	0.1 M NBu_4PF_6 in ACN	+0.27 V vs. Ag/AgCl	Glass	Zn-porphyrin layers on ITO electrode and ITO modified with the poly-1	Pt (10)	[128]
TTF	0.1 M NBu_4PF_6 in ACN	+0.35 V vs. Ag/AgCl	Glass	Zn-porphyrin layers on ITO electrode and ITO modified with the poly-1	Pt (10)	[128]
$[Ru(NH_3)_6]^{3+}$	1 mM $Ru(NH_3)_6Cl_3$ + 0.1 M KCl	−0.35 V vs. Ag/AgCl/KCl sat.	Pt and SS	Amorphous alumina thin film by the MOCVD process	Pt (25)	[115]
$[Ru(NH_3)_6]^{3+}$	1 mM $Ru(NH_3)_6Cl_3$ + 0.1 M Na_2SO_4	-	Silicon wafers	Pt/Al_2O_3 samples	Pt (25)	[174]
$[Ru(NH_3)_6]^{3+}$	1 mM $Ru(NH_3)_6Cl_3$ + 0.1 M Na_2SO_4	-	SS	Pt-TiO_2 prepared by the MOCVD procedure	Pt (25)	[162]
$[Ru(NH_3)_6]^{3+}$	1 mM $Ru(NH_3)_6Cl_3$ + 0.1 M KCl	−0.35 V vs. Ag/AgCl/KCl sat.	Pt wire	Pt/Al_2O_3	Pt (5 to 25)	[175]
$[Ru(NH_3)_6]^{3+}$	1 mM $Ru(NH_3)_6Cl_3$ + 0.1 M KCl	+0.2 V vs. Ag/AgCl	Steel	Polyester paint	Pt (25)	[176]
$[Ru(NH_3)_6]^{3+}$	0.1 M KCl	−0.4 V vs. Ag/AgCl	GCEs	Bismuth film	Pt (25)	[156]
$[Ru(NH_3)_6]^{3+}$	1 mM $Ru(NH_3)_6{}^{3+}$ + 0.1 M KCl	-	-	PTFE	BDD (6 to 23)	[53]
$[Ru(NH_3)_6]^{3+}$	1 mM $Ru(NH_3)_6Cl_3$ + 0.1 M KCl	−0.4 mV vs. Ag/AgCl	Glass	PES and PES-PPy/PW12O40	Pt (100)	[177]
$[Ru(NH_3)_6]^{3+}$	5 mM $Ru(NH_3)_6Cl_3$ + 0.1 M KCl K_2SO_4	−0.70 V vs. Ag/AgCl	NiTi	Electrodeposited tantalum layer	Pt (10)	[178]
$[Ru(NH_3)_6]^{3+}$	0.01 M $Ru(NH_3)_6Cl_3$ + 0.1 M Na_2SO_4 or NaCl	−0.3 V vs. Ag/AgCl		PES, PES–PANI/HSO_4^- and PES–PANI/Cl^-	Pt (100)	[179]
$[Ru(NH_3)_6]^{3+}$	0.01 M $Ru(NH_3)_6Cl_3$ + 0.1 M KCl	−0.4 V vs. Ag/AgCl	PES and PES-PPy/AQSA	PES and PES-PPy/AQSA	Pt (100)	[180]
$[Ru(NH_3)_6]^{3+}$	0.01 M [$Ru(NH_3)_6Cl_3$, $K_3Fe(CN)_6$ and $K_4Fe(CN)_6$] in 0.1 M KCl	−0.4 V vs. Ag/AgCl/KCl (3 M)	-	Polyester fabrics coated with Reduced graphene oxide (RGO)	Pt (100)	[160]
$[Ru(NH_3)_6]^{3+}$	0.01 M $Ru(NH_3)_6Cl_3$ + 0.1 M KCl	−0.4 V vs. Ag/AgCl	-	PES, PES-PPy/$PW_{12}O_{40}{}^{3-}$ and PES-PPy/$PW_{12}O_{40}{}^{3-}$ + PANI	Pt (100)	[181]
$[Ru(NH_3)_6]^{3+}$	0.01 M $Ru(NH_3)_6Cl_3$ + 0.1 M KCl	−0.4 V v. Ag/AgCl	-	RGO	Pt (100)	[182]
$[Ru(NH_3)_6]^{3+}$	0.01 M $Ru(NH_3)_6Cl_3$ + 0.1 M KCl	−0.4 V vs. Ag/AgCl/KCl (3 M)	-	PES-PPy/GO (10%, 20% and 30%)	Pt (100)	[51]
$[Ru(NH_3)_6]^{3+}$	01 mM $Ru(NH_3)_6Cl_3$ + 0.1 M KCl	−0.35 V vs. Ag/AgCl	Pt	Alumina	Pt (25)	[183]

Since some of these mediators were monitored at specific redox potential values (e.g., O_2 or iodine), Table 5 includes that information together with the tip composition and size, whereas the applications are described in terms of the coating or surface film, test environment, and substrate compositions. Often SECM measurements are carried out in aerated chloride-containing electrolytes as a supporting solution due to their aggressive nature toward metals, typically in concentrations close to 3.5 wt.% or 0.1 M. However, simulated biological fluids (SBF) like Ringer's solution have been employed in experiments involving alloys used in biomedical implants [116].

3.2. Generation-Collection Modes

The term generation/collection mode encompasses two different modes of amperometric operations in SECM: tip generation/substrate collection (TG/SC) (see Figure 6A,B), and substrate generation/tip collection (SG/TC) (in Figure 6C), the main difference being the site at which the redox reaction employed for imaging occurs; either at the substrate or at the tip [20]. Although the tip and the substrate both act as working electrodes, the corrosion processes at the substrate are sufficient to develop a spontaneous potential that sustains the reaction without the need to polarize the substrate [184]. In this case, only the application of a potential to the SECM tip is necessary to measure the current flowing at the tip. An alternative situation occurs when the bipotentiostat is employed to set the substrate potential as well as the tip, since the instrument can be used to measure current in both the SECM tip and the substrate.

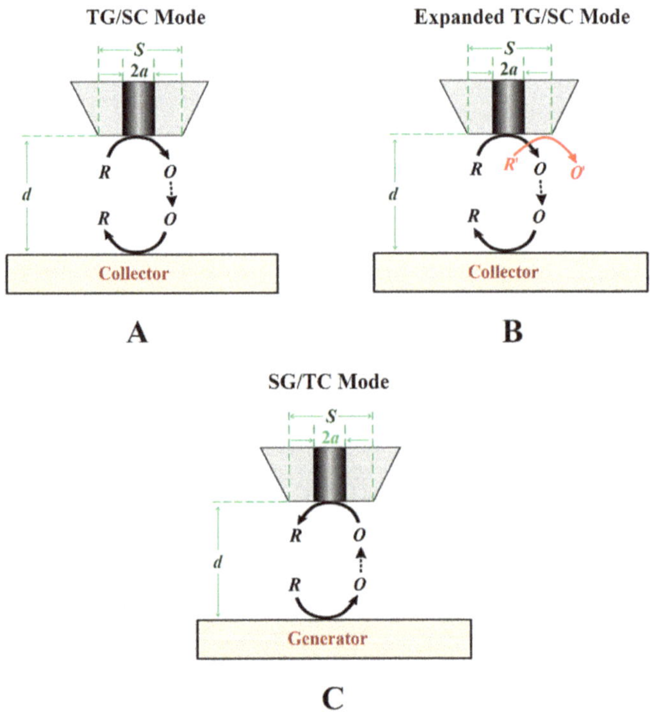

Figure 6. Schemes of generation-collection (G/C) modes of SECM. Types of G/C: (**A**) conventional tip generation and substrate collection, (**B**) bi-reaction tip generation and substrate collection, and (**C**) sample generation and tip collection.

In the TG/SC mode, the electroactive species that can be detected at the substrate is generated at the tip. In this case, the tip and the substrate must have different potentials, either using the bipotentiostat or by the substrate developing a different open circuit potential (OCP). A multireaction TG/SC mode was introduced by Leonard and Bard [185] in 2013 by using the redox conversion of two different species at the tip, as shown in Figure 6B. Depending on the potential required to reduce *O* to *R* or *O′* to *R′* at the substrate, if it is different for both reactions and a total collection efficiency of *O* from *R* can be set at the tip, then the current associated with each of the two reduction reactions can be separated. In the specific case of corrosion systems, other possibilities for bi-reaction interfaces can include two oxidation reactions or a reduction combined with an oxidation.

In SG/TC, the current at the microelectrode arises from a species generated at the surface of the substrate (Figure 6C). This is the traditional G/C mode, and it has an important application for determining the reaction rates in function of the tip-to-substrate distance. If *R* reacts during its transport from the tip to the substrate, the relation between the current intensity at the substrate and at the tip becomes smaller and will greatly vary with the distance, *d*, and it can be used to obtain the rate constant of the homogeneous reaction.

In the G/C mode, the most frequently encountered situation is that redox mediators are selected among the species generated at the substrate under study under active corrosion conditions. The detection of the chemical species involved in the metallic corrosion process allows obtaining concentration profiles in the adjacent electrolyte to the sample. According to Table 6, five mediators have been employed, with the Fe ion as the most used (62% of the cases), followed by the hydrogen ion.

Table 6. Redox mediators used in generation-collection mode SECM for the characterization of thin surface layers and coatings on metals.

Mediator	Electrolyte Solution	E_{tip}	Substrate	Coating	Tip (φ, μm)	Reference
Fe	5–7 mM $FeSO_4(NH_4)_2SO_4$ + 0.1 M KCl	+0.77 V and +0.60 V vs. Ag/AgCl/KCl sat.	CS	Two-component polyurethane	Pt (10)	[66,166]
Fe	5 mM FcMeOH + 0.05 M	+0.6 V vs. Ag/AgCl/KCl sat.	AA2024-T3	Epoxy coating and epoxy coating containing silylester doped capsules	Pt (5)	[132]
Fe	3.5 wt.% NaCl	+0.60 V vs. Ag/AgCl	CS	Epoxy	Pt (10)	[186]
Fe	0.1 M NaCl	+0.60 V vs. Ag/AgCl, KCl sat.	HT steel	Epoxy	Pt (10)	[187]
Fe	1 M $NaClO_4$ and 1 mM $HClO_4$	+0.5 V vs. Ag/AgCl/KCl (3 M)	Steel	Two-component epoxy	Pt (10)	[188]
Fe	0.1 M KCl	+0.6 V vs. Ag/AgCl/KCl sat.	MS	Polyurethane	Pt (10)	[170]
Fe	0.1 M NaCl	+0.6 V vs. Ag/AgCl/KCl sat.	Steel	CrN and TiN PVD	Pt (10)	[189]
Fe	3.5 wt.% NaCl	+0.6 V vs. Ag/AgCl	MS	Epoxy + WO_3 nanoparticle	Pt (10)	[190]
Fe	Natural seawater	+0.3 V vs. Ag/AgCl	MS	DGEBA + CeO_2 nanoparticles	Pt (10)	[191]
Fe	Natural seawater	+0.60 V vs. Ag/AgCl	MS	Neat epoxy and epoxy-(3-aminopropyl)triethoxysilane (APTES) modified MoO_3 nanocomposite	Pt (10)	[192]
Fe	0.1 M NaCl	+0.60 V vs. Ag/AgCl/KCl sat.	CS	Epoxy coatings containing magnesium nanoparticles	Pt (10)	[193]

Table 6. *Cont.*

Mediator	Electrolyte Solution	E_{tip}	Substrate	Coating	Tip (φ, μm)	Reference
Fe	0.1 M NaCl	+0.60 V v Ag/AgCl/KCl sat.	Mn steel	Epoxy	Pt (10)	[194]
H_2O_2	0.1 M KCl	-	MS	Polyurethane	Pt (10)	[170]
H_2O_2	0.1 M KCl	+0.25 V vs. Ag/AgCl/KCl sat.	CS	Polyurethane	Pt (10)	[66]
O_2	0.1 M KCl	+0.4 V and +0.7 V vs. Ag/AgCl/KCl sat.	CS	Two-component polyurethane	Pt (10)	[166]
H_2	SBF	0.0 V vs. Ag/AgCl/KCl (3 M)	AZNd Mg alloy	$Mg(OH)_2$ passive layer	Pt (25)	[195]
H_2	1 mM $Pr(NO_3)_3$ + SBF	0.0 V vs. Ag/AgCl	AZNd Mg alloy	Praseodymium conversion layers	Pt (25)	[196]
H_2	0.01 M NaCl	−0.05 V vs. Ag/AgCl-QRE	AZ31B magnesium alloy	PEDOT	Pt/IrOx (25)	[48]
H_2	SBF	0.0 V vs. Ag/AgCl	AZ31 magnesium alloy	Coating induced by phosphate-based ionic liquids	Pt (10)	[197]
$Ru(NH_3)_6]^{3+}$	5 mM $Ru(NH_3)_6Cl_3$ + 0.1 M K_2SO_4	+0.1 V vs. Ag/AgCl	NiTi	Electrodeposited tantalum layer	Pt (10)	[178]

3.3. Redox Competition Mode

The redox competition mode was introduced by Schuhmann and co-workers in 2006 [198] in a work related to catalysis. In this mode, the SECM tip and the substrate are polarized using the bipotentiostat. When the tip and the substrate are close to each other, they compete for the same redox species (see Figure 7), although the current is measured only at the tip. In a typical corrosion system formed by a metal covered by a non-conductive coating, the oxygen reduction current measured at the SECM tip remains constant while the tip is scanned over the non-defective coating [199]. However, if a scratch is made through the coating system and the metal is exposed to the solution, the current measured at the SECM tip will decrease as the tip explores this active area because the redox species (e.g., oxygen) is consumed at both working electrodes [152,200], and the decrease observed can be correlated to the chemical activity of the substrate that corrodes after its direct exposure to the aggressive environment.

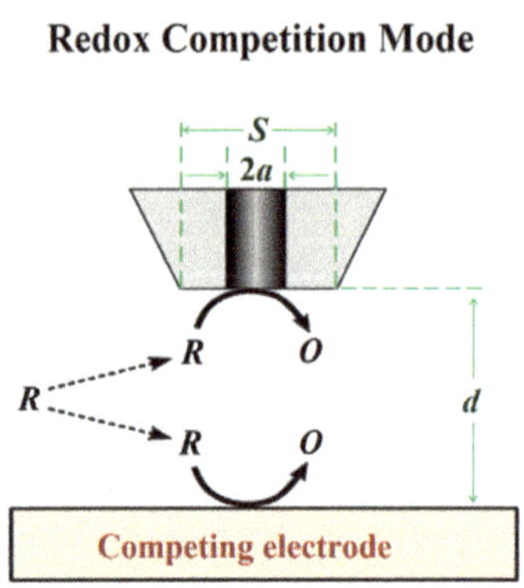

Figure 7. Scheme of the redox competition mode of SECM.

In this mode, dissolved molecular O_2 has been used as a redox mediator in all cases, as shown in Table 7. The most usual application consists in producing a scratch to the coating in order to allow the exposure of the underlying metal to the environment. With this operation mode and using O_2 as mediator, the competition between the O_2 consumed at the tip and that at the substrate (related to the corrosion reaction) can be monitored, even for non biased substrates [1].

Table 7. Redox mediators used in redox competition mode SECM for the characterization of thin surface layers and coatings on metals.

Mediator	Electrolyte Solution	E_{tip}	Substrate	Coating	Tip (φ, μm)	Reference
O_2	0.5 mM FcMeOH + 0.1 M NaCl	−0.6 V vs. Ag/AgCl	AA2024 aluminum alloy	1% γ-APS-doped and 2.5% BTSE-doped epoxy coating	Pt (25)	[129]
O_2	0.1 M KCl	−0.70 V vs. Ag/AgCl/KCl sat.	CS	Two-component polyurethane film	Pt (10)	[166]
O_2	0.9 mM FcMeOH + 5 wt.% NaCl	−0.7 V vs. Ag/AgCl/KCl sat.	Q235 MS	Enamel coating	-	[130]
O_2	0.5 mM FcMeOH + 3.5 wt.% NaCl	−0.7 V vs. Ag/AgCl/KCl sat.	High strength steel (SAPH440)	Enamel coating	-	[131]
O_2	5 mM FcMeOH + 0.05 M NaCl	−0.6 V vs. Ag/AgCl/KCl sat.	AA2024-T3	Epoxy coating and epoxy coating containing silyl-ester doped capsules	Pt (5)	[132]
O_2	1 mM $K_4[Fe(CN)_6]$ + 0.1 M NaCl	−0.7 V vs. Ag/AgCl/KCl sat.	GS	Two-component epoxy primer containing zinc phosphate	Pt (10)	[152]
O_2	3 wt.% NaCl	−0.65 V vs. Ag/AgCl/KCl (3 M)	CS	Polypyrrole	Pt (25)	[201]
O_2	0.2 M H_3BO_3 + 0.05 M BBS + 3.5 wt.% NaCl	-0.642 V vs. SCE	304 SS	CrN film	Pt (15)	[202]
O_2	0.05 M NaCl	−0.6 V vs. Ag/AgCl/KCl sat.	AA2024-T3 aluminum alloy	SMPU polymer containing 8% polyurethane	Pt (10)	[203]
O_2	10 mM NaCl	−0.65 V vs. Ag/AgCl/KCl (3 M)	Cu + Fe	BTAH	Pt (25)	[83]
O_2	1 mM FcMeOH + 0.2 M KCl	−0.7 V vs. Ag/AgCl/KCl sat.	Al flakes	PPy doped with either tungstate or vanadate	Pt (10)	[204]
O_2	1 mM FcMeOH + 0.2 M KCl	−0.7 V vs. Ag/AgCl	AA2024-T3 aluminum alloy	Epoxy resin + vanadate- and tungstate-doped PPy/Al flake composite pigments	Pt (10)	[139]
O_2	3.5 wt.% NaCl	−0.70 V vs. Ag/AgCl/KCl sat.	CS	Epoxy	Pt (10)	[186]
O_2	0.1 M NaCl	−0.70 V vs. Ag/AgCl/KCl sat.	High tensile strength (HT) steel	Epoxy	Pt (10)	[187]

Table 7. *Cont.*

Mediator	Electrolyte Solution	E_{tip}	Substrate	Coating	Tip (φ, μm)	Reference
O_2	3.5 wt.% NaCl	−0.75 V vs. SCE	Q235 MS	PANI + TiO_2 particles	Pt (10)	[205]
O_2	0.1 M NaCl	−0.70 V vs. Ag/AgCl	CS	Epoxy	Pt (10)	[206]
O_2	1 M $NaClO_4$ + 1 mM $HClO_4$	−0.7 V vs. Ag/AgCl/KCl (3 M)	Steel	Two-component epoxy	Pt (10)	[188]
O_2	3.5 wt.% NaCl	−0.75 V vs. SCE	Q235 CS	BTAH and a SMP	Pt (10)	[207]
O_2	3.5 wt.% NaCl	−0.7 V vs. Ag/AgCl/KCl (3 M)	Low-CS	NFC and MFC	Pt (10)	[208]
O_2	0.1 wt.% NaCl + DHS	−0.75 V vs. Ag/AgCl	AA 2024-T3 Aluminium alloy	Silane-modified multi-layer with Mg-rich pigment	Pt (10)	[209]
O_2	DHS	−0.75 V vs. Ag/AgCl	AA 2024-T3 Aluminium alloy	Silane-modified multi-layer with Mg-rich pigment	Pt (10)	[210]
O_2	0.1 M KCl	−0.70 V vs. Ag/AgCl/KCl sat.	MS	Polyurethane	Pt (10)	[170]
O_2	0.1 M KCl + 0.1 M $Na_2B_4O_7$	−0.6 V vs. Ag/AgCl/KCl sat.	CS	Polyurethane	Pt (10)	[199]
O_2	3.5 wt.% NaCl	−0.75 V vs. SCE	Q235 CS	Epoxy	Pt (25)	[211]
O_2	Natural seawater	−0.70 V vs. Ag/AgCl	MS	DGEBA + CeO_2 nanoparticles	Pt (10)	[191]
O_2	0.1 M NaCl	−0.70 V vs. Ag/AgCl/KCl sat.	CS	Epoxy coatings containing magnesium nanoparticles	Pt (10)	[193]
O_2	5% NaCl	−0.70 V vs. Ag/AgCl/KCl sat.	CS	Zinc coating with SiO_2 nanoparticles	Pt (10)	[200]
O_2	0.5 mM FcMeOH + 5 wt.% NaCl	−0.70 V vs. Ag/AgCl/KCl sat.	CS	Silane-Zinc	Pt (10)	[212]
O_2	0.1 M NaCl	−0.70 V vs. Ag/AgCl/KCl sat.	Mn steel	Epoxy	Pt (10)	[194]
O_2	3.5 wt.% NaCl	−0.75 V vs. SCE	Q235 CS	Waterborne epoxy resin + ATP nanoparticles	-	[213]
O_2	0.1 M KCl, 0.1 M Na_2SO_4 and 0.1 M $Na_2B_4O_7$	−0.60 V vs. Ag/AgCl/KCl sat.	CS	Two-component epoxy-polyamine film containing glass flake	Pt (10)	[214]
O_2	3.5 wt.% NaCl	−0.7 V vs. Ag/AgCl/KCl sat.	7075 aluminum alloy	Berberine	Pt (10)	[172]
O_2	3.5 wt.% NaCl	−0.75 V vs. SCE	AA2024-T3 aluminum alloy	Shape memory epoxy polymers containing dual-function microspheres	Pt (−)	[215]

3.4. Combined Operation Modes

The signal recorded by the sensing probes of SECM consists of a complex combination of spatially-resolved information originating from the distance between the tip and the surface of the sample under study (i.e., sensitive to the morphology of the substrate) and of the actual chemical response due to the reactivity of the substrate, which in practice mainly limits its use to the characterization of flat surfaces and to the first stages of the formation of the surface film due to its progressive rougheness with degradation. Although the morphological and chemical information may be ultimately convoluted in conventional

SECM analysis, the first contribution can be considered constant in the case of a flat surface, so that changes in probe response can be attributed to the chemical reactivity of the studied system. Unfortunately, highly reactive systems occurring in light-weight alloy materials, which rapidly develop layers of oxide products and gas evolution under common operating conditions, and those associated with the self-healing mechanisms of smart coatings containing nanoreservoirs for functionalized operation, do not exhibit flat surfaces. This feature is not a real limitation for the investigation of thin films on metals because their width dimensions are often much smaller than the size of the scanning probe, but it can make it difficult to characterize larger surface defects such as crevices or heterogeneous regions extending over a large surface compared to the tip dimensions such as those formed in welds. Notwithstanding, efficient measurement strategies have been developed to construct SECM surface images by combining separate images of smaller regions [216].

To overcome this limitation imposed by the convolution of topographic information and chemical activity in the signal measured at the tip, a multi-scale electrochemical methodological procedure can be performed to deconvolute chemical information relevant to corrosion reactions and protection mechanisms in such complex systems can be performed [217]. Souto and coworkers developed a quite simple and systematic methodological procedure involving the combined use of various operation modes in amperometric SECM to study defects in organic coatings of width and depth dimensions greater than those of the tip [170], as illustrated in Figure 8.

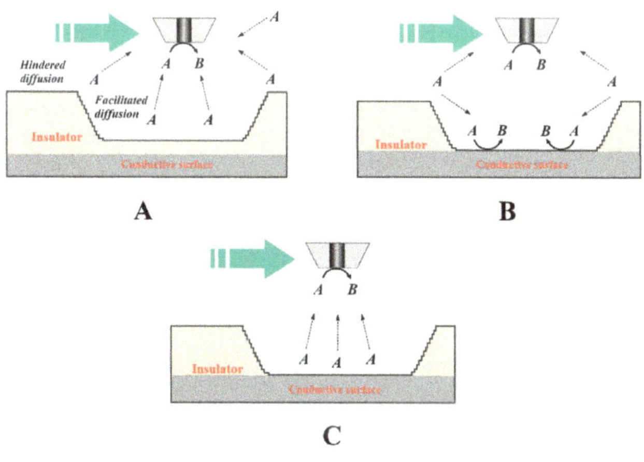

Figure 8. Diagrams of the processes that occur at the tip for the electroreduction of species *A* and *B* when the tip passes over a larger defect depending on whether it is an insulating or a conductive surface exposed to the aqueous medium. The green arrow indicates the scan direction. Depending on the source of species *A*, the following situations have been described: (**A**) species *A* is present in the aqueous medium and is not consumed within the defect, which behaves as an insulator; (**B**) species *A* is present in the aqueous medium and transforms at both the tip and the bottom of the defect; and (**C**) species *A* is generated at the bottom of the defect, although it was not originally present in the aqueous environment.

This procedure required the choice of detection modes more sensitive to either the topographical changes or the chemical activity by controlling the local chemistry of the system and the characteristics of the tip, followed by a subsequent stage of recording the combined signal of the complete corroding system. This methodology can be further extended by combining potentiometric modes with amperometric operation using multi-probe configurations, as recently demonstrated for the study of corrosion processes on cut edges of organically coated galvanized steel [42]. In this case, the combined amperometric/potentiometric SECM operation was performed by fabricating a multi-probe assembly using the same procedures previously developed for fabricating potentiometric probes with an internal reference electrode [40,120].

An alternate way to overcome the referred limitation is to associate SECM with other surface resolved techniques [218]. An option is the combination of SECM with AFM using cantilever probes modified for this purpose, which made it possible to simultaneously image the topography and the electrochemical activity in situ. In this way, the monitoring of nucleating corrosion pits in iron-based materials [219] and the dissolution-redeposition of metal ions in acidic environment [220] have been succesfully imaged in situ.

Besides the combination of SECM with topographically sensitive techniques, software routines can be designed so that the scan is actually performed at a constant tip-substrate distance (i.e., following the actual surface of the sample) instead of operating at a constant height, as is normally done in convential SECM operation [221]. Alternately, the measurement of shear forces between the tip and the surface can be used for constant distance operation in SECM instead of AFM [221]. The success of such an association was reported by Etienne et al. for monitoring the performance of self-healing coatings deposited on an aluminium alloy [222]. In their work, local features with a depth profile in excess of 50 μm were successfully resolved. In addition, local chemical analysis was simultaneously performed using in situ Raman spectroscopy.

Finally, corrosion reactions that progress far beyond their initial stages and eventually reach dimensions of a hundred micrometers or a few millimeters are already accessible using other experimental techniques and would not require the micrometric resolution of SECM.

4. Concluding Remarks

SECM is a powerful technique for the analysis and characterization of coatings of different nature applied over conductive and non-conductive substrates, as well as for self-healing coatings.

Three main operation modes are employed in amperometric SECM to this end, namely the feedback (both positive and negative feedback), generation/collection (G/C) mode (in either SG/TC or TG/SC configuration), and the redox competition modes. In some cases, various operation modes are used in the same experiment. Feedback mode is the most usual procedure, mainly due to the need of a controlled approach of the tip to the substrate.

The selection of the mediator depends on several factors, including the operation mode or the nature of the substrate and the solution employed. The most frequently employed mediators are ferrocene-methanol and hexacyanoferrate (III) in the feedback operation, iron in the G/C mode, and dissolved O_2 in redox competition mode. Pt disks of 10 and 25 μm of diameter are usually employed as tips. In minor proportion, Pt coated with parylene C, Pt/IrOx, Au and boron doped diamond tips are also used.

A rather recent addition to the characterization of surface layers and coatings applied on metals is the use of AC signals by using active MEs such as those employed in conventional amperometric operation of the SECM. The higher spatial resolution of the method, especially in the case of thin surface layers, makes these modes very attractive, in addition to the power of recording the images at different frequencies, thus allowing different processes to be distinguished in one measurement. Furthermore, by extracting the electrochemical impedance from the localised potential and current signals, a frequency-resolved scanning electrochemical impedance microscopy (SEIM) technique is available.

Enhanced chemical selectivity is obtained by using micropotentiometric sensors in the SECM. Although their application was initially hindered by the significantly slower scanning rates that can be used due to the finite response times of ISMEs, which often compares poorly with the dynamics of the corrosion process under investigation, the development of new scanning and deconvolution methods are making this operation more attractive for the characterization of thin surface layers, as it already has with polymer coatings.

Author Contributions: Conceptualization, J.J.S.; investigation, J.J.S., J.I. and R.M.S.; data curation, J.J.S. and R.M.S.; methodology, J.I. and R.M.S.; writing—original draft preparation, J.J.S. and R.M.S.; writing—review and editing, J.J.S., J.I. and R.M.S.; visualization, J.J.S. and R.M.S.; funding acquisition, R.M.S. All authors have read and agreed to the published version of the manuscript.

Funding: The work was supported by the University of La Laguna and the Spanish Ministry of Science, Innovation and Universities (Madrid, Spain) under contract No. 2022/0000586.

Institutional Review Board Statement: Not applicable.

Informed Consent Statement: Not applicable.

Data Availability Statement: Not applicable.

Acknowledgments: J.J.S. acknowledges Géza Nagy for hosting his scientific mobility work to the University of Pécs (Hungary) from September to December, 2019, during which this work was designed and initiated.

Conflicts of Interest: The authors declare no conflict of interest.

References

1. González-García, Y.; Santana, J.J.; González-Guzmán, J.; Izquierdo, J.; González, S.; Souto, R.M. Scanning electrochemical microscopy for the investigation of localized degradation processes in coated metals. *Prog. Org. Coat.* **2010**, *69*, 110–117. [CrossRef]
2. Blaiszik, B.J.; Kramer, S.L.B.; Olugebefola, S.C.; Moore, J.S.; Sottos, N.R.; White, S.R. Self-healing polymers and composites. *Annu. Rev. Mater.* **2010**, *40*, 179–211. [CrossRef]
3. Stankiewicz, A.; Szczygieł, I.; Szczygieł, B. Self-healing coatings in anti-corrosion applications. *J. Mater. Sci.* **2013**, *48*, 8041–8051. [CrossRef]
4. Montemor, M.F. Functional and smart coatings for corrosion protection: A review of recent advances. *Surf. Coat. Technol.* **2014**, *258*, 17–37. [CrossRef]
5. Pistorius, P.C.; Burstein, G.T. Metastable pitting corrosion of stainless steel and the transition to stability. *Phil. Trans. Roy. Soc. London A* **1992**, *341*, 531–559.
6. Tan, Y. Sensing localised corrosion by means of electrochemical noise detection and analysis. *Sens. Actuator B-Chem.* **2010**, *139*, 688–698. [CrossRef]
7. Suter, T.; Böhni, H. A new microelectrochemical method to study pit initiation on stainless steels. *Electrochim. Acta* **1997**, *42*, 3275–3280. [CrossRef]
8. Gupta, R.K.; Sukiman, N.L.; Cavanaugh, M.K.; Hinton, B.R.W.; Hutchinson, C.R.; Birbilis, N. Metastable pitting characteristics of aluminium alloys measured using current transients during potentiostatic polarisation. *Electrochim. Acta* **2012**, *66*, 245–254. [CrossRef]
9. Martin, F.A.; Bataillon, C.; Cousty, J. In situ AFM detection of pit onset location on a 304L stainless steel. *Corros. Sci.* **2008**, *50*, 84–92. [CrossRef]
10. Birbilis, N.; Meyer, K.; Muddle, B.C.; Lynch, S.P. In situ measurement of corrosion on the nanoscale. *Corros. Sci.* **2009**, *51*, 1569–1572. [CrossRef]
11. Maurice, V.; Marcus, P. Passive films at the nanoscale. *Electrochim. Acta* **2012**, *84*, 129–138. [CrossRef]
12. Lillard, R.S. Scanning electrode techniques for investigating near-surface solution current densities. In *Analytical Methods in Corrosion Science and Engineering*; Marcus, P., Mansfeld, F., Eds.; CRC Press: Boca Raton, FL, USA, 2006; pp. 571–604.
13. Jensen, M.B.; Tallman, D.E. Application of SECM to corrosion studies. In *Electroanalytical Chemistry: A Series of Advances*; Bard, A.J., Zoski, C., Eds.; CRC Press: Boca Raton, FL, USA, 2012; Volume 24, pp. 171–286.
14. Payne, N.A.; Stephens, L.I.; Mauzeroll, J. The application of scanning electrochemical microscopy to corrosion research. *Corrosion* **2017**, *73*, 759–780. [CrossRef]
15. Huang, V.M.-W.; Wu, S.–H.; Orazem, M.E.; Pébère, N.; Tribollet, B.; Vivier, V. Local electrochemical impedance spectroscopy: A review and some recent developments. *Electrochim. Acta* **2011**, *56*, 8048–8057. [CrossRef]
16. Engstrom, R.C.; Pharr, C.M. Scanning electrochemical microscopy. *Anal. Chem.* **1989**, *61*, 1099A–1104A. [CrossRef]
17. Bard, A.J.; Fan, F.R.F.; Kwak, J.; Lev, O. Scanning electrochemical microscopy. Introduction and principles. *Anal. Chem.* **1989**, *61*, 132–138. [CrossRef]
18. Polcari, D.; Dauphin-Ducharme, P.; Mauzeroll, J. Scanning electrochemical microscopy: A comprehensive review of experimental parameters from 1989 to 2015. *Chem. Rev.* **2016**, *116*, 13234–13278. [CrossRef]
19. Farrokhpay, S. Application of spectroscopy and microscopy techniques in surface coatings evaluation: A review. *Appl. Spectrosc. Rev.* **2012**, *47*, 233–243. [CrossRef]
20. Zoski, C.G. Review-Advances in scanning electrochemical microscopy (SECM). *J. Electrochem. Soc.* **2016**, *163*, H3088–H3100. [CrossRef]

21. Izquierdo, J.; Knittel, P.; Kranz, C. Scanning electrochemical microscopy: An analytical perspective. *Anal. Bioanal. Chem.* **2018**, *410*, 307–324. [CrossRef]
22. Etienne, M.; Schulte, A.; Schuhmann, W. High resolution constant-distance mode alternating current scanning electrochemical microscopy (AC-SECM). *Electrochem. Commun.* **2004**, *6*, 288–293. [CrossRef]
23. Kuznetsov, V.; Maljusch, A.; Souto, R.M.; Bandarenka, A.S.; Schuhmann, W. Characterisation of localised corrosion processes using scanning electrochemical impedance microscopy. *Electrochem. Commun.* **2014**, *4*, 38–41. [CrossRef]
24. Katemann, B.B.; Inchauspe, C.G.; Castro, P.A.; Schulte, A.; Calvo, E.J.; Schuhmann, W. Precursor sites for localised corrosion on lacquered tinplates visualised by means of alternating current scanning electrochemical microscopy. *Electrochim. Acta* **2003**, *48*, 1115–1121. [CrossRef]
25. Eckhard, K.; Erichsen, T.; Stratmann, M.; Schuhmann, W. Frequency-dependent alternating-current scanning electrochemical microscopy (4D AC-SECM) for local visualisation of corrosion sites. *Chem. Eur. J.* **2008**, *14*, 3968–3976. [CrossRef] [PubMed]
26. Eckhard, K.; Schuhmann, W.; Maciejewska, M. Determination of optimum imaging conditions in AC-SECM using the mathematical distance between approach curves displayed in the impedance domain. *Electrochim. Acta* **2009**, *54*, 2125–2130. [CrossRef]
27. Pähler, M.; Santana, J.J.; Schuhmann, W.; Souto, R.M. Application of AC-SECM in Corrosion Science—Local visualisation of inhibitor films on active metals for corrosion protection. *Chem. Eur. J.* **2011**, *17*, 905–911. [CrossRef]
28. Pähler, M.; Santana, J.J.; Schuhmann, W.; Souto, R.M. Investigation of copper corrosion inhibition with frequency-dependent alternating-current scanning electrochemical microscopy. *ChemPlusChem* **2012**, *77*, 707–712.
29. Souto, R.M.; Socas, B.; Izquierdo, J.; Santana, J.J.; González, S. New opportunities for the study of organic films applied on metals for corrosion protection by means of alternating current scanning electrochemical microscopy. *Prog. Org. Coat.* **2012**, *74*, 371–375. [CrossRef]
30. Santana, J.J.; Pähler, M.; Souto, R.M.; Schuhmann, W. Direct evidence of early blister formation in polymer-coated metals from exposure to chloride-containing electrolytes by alternating-current scanning electrochemical microscopy. *Electrochim. Acta* **2012**, *77*, 60–64. [CrossRef]
31. Bandarenka, A.S.; Eckhard, K.; Maljusch, A.; Schuhmann, W. Localized electrochemical impedance spectroscopy: Visualization of spatial distributions of the key parameters describing solid/liquid interfaces. *Anal. Chem.* **2013**, *85*, 2443–2448. [CrossRef]
32. Bandarenka, A.S.; Maljusch, A.; Kuznetsov, V.; Eckhard, K.; Schuhmann, W. Localized electrochemical impedance measurements for electrochemical surface science. *J. Phys. Chem. C* **2014**, *118*, 8952–8959. [CrossRef]
33. Kuznetsov, V.; Estrada-Vargas, A.; Maljusch, A.; Berkes, B.B.; Bandarenka, A.S.; Souto, R.M.; Schuhmann, W. Kinetic passivation effect of localized differential aeration on brass. *ChemPlusChem* **2016**, *81*, 49–57. [CrossRef] [PubMed]
34. Estrada-Vargas, A.; Bandarenka, A.; Kuznetsov, V.; Schuhmann, W. In situ characterization of ultrathin films by scanning electrochemical impedance spectroscopy. *Anal. Chem.* **2016**, *88*, 3354–3362. [CrossRef] [PubMed]
35. Wang, W.; Li, W.; Song, L.; Fan, W.; Xiong, C.; Gao, X.; Zhang, X.; Liu, X. Self-healing performance of coatings containing synthetic hexamethylene diisocyanate biuret microcapsules. *J. Electrochem. Soc.* **2017**, *164*, C635–C640. [CrossRef]
36. Wang, W.; Wang, H.; Zhao, J.; Wang, X.; Xiong, C.; Song, L.; Ding, R.; Han, P.; Li, W. Self-healing performance and corrosion resistance of graphene oxide-mesoporous silicon layer-nanosphere structure coating under marine alternating hydrostatic pressure. *Chem. Eng. J.* **2019**, *361*, 792–804. [CrossRef]
37. Lamaka, S.V.; Souto, R.M.; Ferreira, M.G.S. In-situ visualization of local corrosion by scanning ion-selective electrode technique (SIET). In *Microscopy: Science, Technology, Applications and Education*; Méndez-Vilas, A., Díaz, J., Eds.; Formatex Research Center: Badajoz, Spain, 2010; Volume 3, pp. 2162–2173.
38. Kiss, A.; Filotás, D.; Souto, R.M.; Nagy, G. The effect of electric field on potentiometric scanning electrochemical microscopic imaging. *Electrochem. Commun.* **2017**, *77*, 138–141. [CrossRef]
39. Filotás, D.; Fernández-Pérez, B.M.; Izquierdo, J.; Kiss, A.; Nagy, L.; Nagy, G.; Souto, R.M. Improved potentiometric SECM imaging of galvanic corrosion reactions. *Corros. Sci.* **2017**, *129*, 136–145. [CrossRef]
40. Filotás, D.; Fernández-Pérez, B.M.; Kiss, A.; Nagy, L.; Nagy, G.; Souto, R.M. Double barrel microelectrode assembly to prevent electrical field effects in potentiometric SECM imaging of galvanic corrosion processes. *J. Electrochem. Soc.* **2018**, *165*, C270–C277. [CrossRef]
41. Marques, A.G.; Izquierdo, J.; Souto, R.M.; Simões, A.M. SECM imaging of the cut edge corrosion of galvanized steel as a function of pH. *Electrochim. Acta* **2015**, *153*, 238–245. [CrossRef]
42. Filotás, D.; Izquierdo, J.; Fernández-Pérez, B.M.; Nagy, L.; Nagy, G.; Souto, R.M. Contributions of microelectrochemical scanning techniques for the efficient detection of localized corrosion processes at the cut edges of polymer-coated galvanized steel. *Molecules* **2022**, *27*, 2167. [CrossRef]
43. Stulík, K.; Amatore, C.; Holub, K.; Marecek, V.; Kutner, W. Microelectrodes. Definitions, characterization, and applications (Technical report). *Pure Appl. Chem.* **2000**, *72*, 1483–1492. [CrossRef]
44. Izquierdo, J.; Nagy, L.; Varga, Á.; Santana, J.J.; Nagy, G.; Souto, R.M. Spatially resolved measurement of electrochemical activity and pH distributions in corrosion processes by scanning electrochemical microscopy using antimony microelectrode tips. *Electrochim. Acta* **2011**, *56*, 8846–8850. [CrossRef]
45. Zhu, Z.J.; Ye, Z.N.; Zhang, J.Q.; Cao, F.H. Novel dual Pt-Pt-IrOx ultramicroelectrode for pH imaging using SECM in both potentiometric and amperometric modes. *Electrochem. Commun.* **2018**, *88*, 47–51. [CrossRef]

46. Lefrou, C.; Cornut, R. Analytical expressions for quantitative scanning electrochemical microscopy (SECM). *Chem. Phys. Chem.* **2010**, *11*, 547–556. [CrossRef] [PubMed]
47. Cortés-Salazar, F.; Deng, H.; Peljo, P.; Pereira, C.M.; Kontturi, K.; Girault, H.H. Parylene C coated microelectrodes for scanning electrochemical microscopy. *Electrochim. Acta* **2013**, *110*, 22–29. [CrossRef]
48. Tefashe, U.M.; Dauphin-Ducharme, P.; Danaie, M.; Cano, Z.P.; Kish, J.R.; Botton, G.A.; Mauzeroll, J. Localized corrosion behavior of AZ31B magnesium alloy with an electrodeposited poly (3,4-ethylenedioxythiophene) coating. *J. Electrochem. Soc.* **2015**, *162*, C536–C544. [CrossRef]
49. Combellas, C.; Kanoufi, F.; Mazouzi, D.; Thiébault, A. Surface modification of halogenated polymers: 5. Localized electroless deposition of metals on poly(tetrafluoroethylene) surfaces. *J. Electroanal. Chem.* **2003**, *556*, 43–52. [CrossRef]
50. Grisotto, F.; Ghorbal, A.; Goyer, C.; Charlier, J.; Palacin, S. Direct SECM localized electrografting of vinylic monomers on a conducting substrate. *Chem. Mater.* **2011**, *23*, 1396–1405. [CrossRef]
51. Molina, J.; Zille, A.; Fernández, J.; Souto, A.P.; Bonastre, J.; Cases, F. Conducting fabrics of polyester coated with polypyrrole and doped with graphene oxide. *Synth. Met.* **2015**, *204*, 110–121. [CrossRef]
52. Borgwarth, K.; Ebling, D.; Heinze, J. Applications of scanning ultra micro electrodes for studies on surface conductivity. *Electrochim. Acta* **1995**, *40*, 1455–1460. [CrossRef]
53. Holt, K.B.; Hu, J.; Foord, J.S. Fabrication of boron-doped diamond ultramicroelectrodes for use in scanning electrochemical microscopy experiments. *Anal. Chem.* **2007**, *79*, 2556–2561. [CrossRef]
54. Varvara, S.; Caniglia, G.; Izquierdo, J.; Bostan, R.; Gaina, L.; Souto, R.M. Multiscale electrochemical analysis of the corrosion control of bronze in simulated acid rain by horse-chestnut (*Aesculus hippocastanum* L.) extract as green inhibitor. *Corros. Sci.* **2020**, *165*, 108381. [CrossRef]
55. Zhou, J.; Campbell, C.; Heller, A.; Bard, A.J. Scanning electrochemical microscopy. 44. Imaging of horseradish peroxidase immobilized on insulating substrates. *Anal. Chem.* **2002**, *74*, 4007–4010. [CrossRef] [PubMed]
56. Zhang, J.; Barker, A.L.; Unwin, P.R. Microelectrochemical studies of charge transfer at the interface between two immiscible electrolyte solutions: Electron transfer from decamethyl ferrocene to aqueous oxidants. *J. Electroanal. Chem.* **2000**, *483*, 95–107. [CrossRef]
57. Seegmiller, J.C.; Buttry, D.A. A SECM study of heterogeneous redox activity at AA2024 surfaces. *J. Electrochem. Soc.* **2003**, *150*, 413–418. [CrossRef]
58. Quinto, M.; Jenekhe, S.A.; Bard, A.J. Polymer films on electrodes. 30. Electrochemistry and scanning electrochemical microscopy characterization of benzimidazolebenzophenanthroline-type ladder (BBL) and semiladder (BBB) polymer films. *Chem. Mater.* **2001**, *13*, 2824–2832. [CrossRef]
59. Sánchez-Sánchez, C.M.; Rodríguez-López, J.; Bard, A.J. Scanning electrochemical microscopy. 60. Quantitative calibration of the SECM substrate generation/tip collection mode and its use for the study of the oxygen reduction mechanism. *Anal. Chem.* **2008**, *80*, 3254–3260. [CrossRef]
60. Mandler, D. A new approach to the high resolution electrodeposition of metals via the feedback mode of the scanning electrochemical microscope. *J. Electrochem. Soc.* **1990**, *137*, 1079–1086. [CrossRef]
61. Lee, C.; Bard, A.J. Scanning electrochemical microscopy. Application to polymer and thin metal oxide films. *Anal. Chem.* **1990**, *62*, 1906–1913. [CrossRef]
62. Horrocks, B.R.; Schmidtke, D.; Heller, A.; Bard, A.J. Scanning electrochemical microscopy. 24. Enzyme ultramicroelectrodes for the measurement of hydrogen peroxide at surfaces. *Anal. Chem.* **1993**, *65*, 3605–3614. [CrossRef]
63. Basame, S.B.; White, H.S. Scanning electrochemical microscopy of metal/metal oxide electrodes. Analysis of spatially localized electron-transfer reactions during oxide growth. *Anal. Chem.* **1999**, *71*, 3166–3170. [CrossRef]
64. Basame, S.B.; White, H.S. Scanning electrochemical microscopy of native titanium oxide films. Mapping the potential dependence of spatially-localized electrochemical reactions. *J. Phys. Chem.* **1995**, *99*, 16430–16435. [CrossRef]
65. Park, H.S.; Leonard, K.C.; Bard, A.J. Surface interrogation scanning electrochemical microscopy (SI-SECM) of photoelectrochemistry at a W/Mo-$BiVO_4$ semiconductor electrode: Quantification of hydroxyl radicals during water oxidation. *J. Phys. Chem. C* **2013**, *117*, 12093–12102. [CrossRef]
66. González, S.; Santana, J.J.; González-García, Y.; Fernández-Mérida, L.; Souto, R.M. Scanning electrochemical microscopy for the investigation of localized degradation processes in coated metals: Effect of oxygen. *Corros. Sci.* **2011**, *53*, 1910–1915. [CrossRef]
67. Selzer, Y.; Turyan, I.; Mandler, D. Studying heterogeneous catalysis by the scanning electrochemical microscope (SECM): The reduction of protons by methyl viologen catalyzed by a platinum surface. *J. Phys. Chem. B* **1999**, *103*, 1509–1517. [CrossRef]
68. Fabre, B.; Bassani, D.M.; Liang, C.K.; Ray, D.; Hui, F.; Hapiot, P. Anthracene and anthracene:C_{60} adduct-terminated monolayers covalently bound to hydrogen-terminated silicon surfaces. *J. Phys. Chem. C* **2011**, *115*, 14786–14796. [CrossRef]
69. Bollo, S.; Jara-Ulloa, P.; Finger, S.; Núñez-Vergara, L.J.; Squella, J.A. Scanning electrochemical microscopy (SECM) study of superoxide generation and its reactivity with 1,4-dihydropyridines. *J. Electroanal. Chem.* **2005**, *577*, 235–242. [CrossRef]
70. Liu, B.; Rotenberg, S.A.; Mirkin, M.V. Scanning electrochemical microscopy of living cells. 4. Mechanistic study of charge transfer reactions in human breast cells. *Anal. Chem.* **2002**, *74*, 6340–6348. [CrossRef]
71. Filotás, D.; Asserghine, A.; Nagy, L.; Nagy, G. Short-term influence of interfering ion activity change on ion-selective micropipette electrode potential: Another factor that can affect the time needed for imaging in potentiometric SECM. *Electrochem. Commun.* **2017**, *77*, 62–64. [CrossRef]

72. Kiss, A.; Nagy, G. New SECM scanning algorithms for improved potentiometric imaging of circularly symmetric targets. *Electrochim. Acta* **2014**, *119*, 169–174. [CrossRef]
73. Kiss, A.; Nagy, G. Deconvolution of potentiometric SECM images recorded with high scan rate. *Electrochim. Acta* **2015**, *163*, 303–309. [CrossRef]
74. Horrocks, B.R.; Mirkin, M.V.; Pierce, D.T.; Bard, A.J.; Nagy, G.; Toth, K. Scanning electrochemical microscopy. 19. Ion-selective potentiometric microscopy. *Anal. Chem.* **1993**, *65*, 1213–1224. [CrossRef]
75. Frateur, I.; Bayet, E.; Keddam, M.; Tribollet, B. Local redox potential measurement. *Electrochem. Commun.* **1999**, *1*, 336–340. [CrossRef]
76. Filotás, D.; Fernández-Pérez, B.M.; Izquierdo, J.; Nagy, L.; Nagy, G.; Souto, R.M. Novel dual microelectrode probe for the simultaneous visualization of local Zn^{2+} and pH distributions in galvanic corrosion processes. *Corros. Sci.* **2017**, *114*, 37–44. [CrossRef]
77. Souto, R.M.; Izquierdo, J.; Santana, J.J.; Kiss, A.; Nagy, L.; Nagy, G. Progress in scanning electrochemical microscopy by coupling potentiometric and amperometric measurement modes. In *Current Microscopy Contributions to Advances in Science and Technology*; Méndez-Vilas, A., Ed.; Formatex Research Center: Badajoz, Spain, 2012; Volume 2, pp. 1407–1415.
78. Santos, C.S.; Lima, A.S.; Battistel, D.; Daniele, S.; Bertotti, M. Fabrication and use of dual-function iridium oxide coated gold SECM tips. An application to pH monitoring above a copper electrode surface during nitrate reduction. *Electroanalysis* **2016**, *28*, 1441–1447. [CrossRef]
79. Zhu, Z.J.; Liu, X.Y.; Ye, Z.N.; Zhang, J.Q.; Cao, F.H.; Zhang, J.X. A fabrication of iridium oxide film pH micro-sensor on Pt ultramicroelectrode and its application on in-situ pH distribution of 316L stainless steel corrosion at open circuit potential. *Sens. Actuator B-Chem.* **2018**, *255*, 1974–1982. [CrossRef]
80. Fernández-Pérez, B.M.; Izquierdo, J.; González, S.; Souto, R.M. Scanning electrochemical microscopy studies for the characterization of localized corrosion reactions at cut edges of coil-coated steel. *J. Solid State Electrochem.* **2014**, *18*, 2983–2992. [CrossRef]
81. Fernández-Pérez, B.M.; Izquierdo, J.; Santana, J.J.; González, S.; Souto, R.M. Scanning electrochemical microscopy studies for the characterization of localized corrosion reactions at cut edges of painted galvanized steel as a function of solution pH. *Int. J. Electrochem. Sci.* **2015**, *10*, 10145–10156.
82. Ramírez-Cano, J.A.; Veleva, L.; Fernández-Pérez, B.M.; Souto, R.M. SVET study of the interaction of 2-mercaptobenzothiazole corrosion inhibitor with Au, Cu and Au-Cu galvanic coupling. *Int. J. Corros. Scale Inhib.* **2017**, *6*, 307–317.
83. Izquierdo, J.; Nagy, L.; Santana, J.J.; Nagy, G.; Souto, R.M. A novel microelectrochemical strategy for the study of corrosion inhibitors employing the scanning vibrating electrode technique and dual potentiometric/amperometric operation in scanning electrochemical microscopy: Application to the study of the cathodic inhibition by benzotriazole of the galvanic corrosion of copper coupled to iron. *Electrochim. Acta* **2011**, *58*, 707–716.
84. Maile, F.J.; Schauer, T.; Eisenbach, C.D. Evaluation of corrosion and protection of coated metals with local ion concentration technique (LICT). *Prog. Org. Coat.* **2000**, *38*, 111–116. [CrossRef]
85. Zhu, Z.; Zhang, Q.; Liu, P.; Zhang, J.; Cao, F. Quasi-simultaneous electrochemical/chemical imaging of local Fe^{2+} and pH distributions on 316L stainless steel surface. *J. Electroanal. Chem.* **2020**, *871*, 114107. [CrossRef]
86. Denuault, G.; Nagy, G.; Toth, K. Potentiometric Probes. In *Scanning Electrochemical Microscopy*, 2nd ed.; Bard, A.J., Mirkin, M.V., Eds.; CRC Press: Boca Raton, FL, USA, 2012; pp. 275–316.
87. Ogle, K.; Baudu, V.; Garrigues, L.; Philippe, X. Localized electrochemical methods applied to cut edge corrosion. *J. Electrochem. Soc.* **2000**, *147*, 3654–3660. [CrossRef]
88. Ogle, K.; Morel, S.; Jacquet, D. Observation of self-healing functions on the cut edge of galvanized steel using SVET and pH microscopy. *J. Electrochem. Soc.* **2006**, *153*, B1–B5. [CrossRef]
89. Karavai, O.V.; Bastos, A.C.; Zheludkevich, M.L.; Taryba, M.G.; Lamaka, S.V.; Ferreira, M.G.S. Localized electrochemical study of corrosion inhibition in microdefects on coated AZ31 magnesium alloy. *Electrochim. Acta* **2010**, *55*, 5401–5406. [CrossRef]
90. Lamaka, S.V.; Karavai, O.V.; Bastos, A.C.; Zheludkevich, M.L.; Ferreira, M.G.S. Monitoring local spatial distribution of Mg^{2+}, pH and ionic currents. *Electrochem. Commun.* **2008**, *10*, 259–262. [CrossRef]
91. Zhang, G.; Jiang, E.; Wu, L.; Tang, A.; Atrens, A.; Pan, F. Active corrosion protection of phosphate loaded PEO/LDHs composite coatings: SIET study. *J. Magnes. Alloy.* **2021**, in press. [CrossRef]
92. Snihirova, D.; Lamaka, S.V.; Taryba, M.; Salak, A.N.; Kallip, S.; Zheludkevich, M.L.; Ferreira, M.G.S.; Montemor, M.F. Hydroxyapatite microparticles as feedback-active reservoirs of corrosion inhibitors. *ACS Appl. Mater. Interfaces* **2010**, *2*, 3011–3022. [CrossRef]
93. Ding, H.; Hihara, L.H. Localized corrosion currents and pH profile over B_4C, SiC, and Al_2O_3 reinforced 6092 aluminum composites: I. In 0.5 M Na_2SO_4 solution. *J. Electrochem. Soc.* **2005**, *152*, B161–B167. [CrossRef]
94. Ding, H.; Hawthorn, G.A.; Hihara, L.H. Inhibitive effect of seawater on the corrosion of particulate-reinforced aluminum-matrix composites and monolithic aluminum alloy. *J. Electrochem. Soc.* **2009**, *156*, C352–C359. [CrossRef]
95. Ding, H.; Hihara, L.H. Galvanic corrosion and localized degradation of aluminum-matrix composites reinforced with silicon particulates. *J. Electrochem. Soc.* **2008**, *155*, C226–C233. [CrossRef]

96. Taryba, M.; Lamaka, S.V.; Snihirova, D.; Ferreira, M.G.S.; Montemor, M.F.; Wijting, W.K.; Toews, S.; Grundmeier, G. The combined use of scanning vibrating electrode technique and micro-potentiometry to assess the self-repair processes in defects on "smart" coatings applied to galvanized steel. *Electrochim. Acta* **2011**, *56*, 4475–4488. [CrossRef]
97. Taryba, M.G.; Montemor, M.F.; Lamaka, S.V. Quasi-simultaneous mapping of local current density, pH and dissolved O_2. *Electroanalysis* **2015**, *27*, 2725–2730. [CrossRef]
98. Marques, A.G.; Taryba, M.G.; Panão, A.S.; Lamaka, S.V.; Simões, A.M. Application of scanning electrode techniques for the evaluation of iron–zinc corrosion in nearly neutral chloride solutions. *Corros. Sci.* **2016**, *104*, 123–131. [CrossRef]
99. Izquierdo, J.; Kiss, A.; Santana, J.J.; Nagy, L.; Bitter, I.; Isaacs, H.S.; Nagy, G.; Souto, R.M. Development of Mg^{2+} ion-selective microelectrodes for potentiometric scanning electrochemical microscopy monitoring of galvanic corrosion processes. *J. Electrochem. Soc.* **2013**, *160*, C451–C459. [CrossRef]
100. Filotás, D.; Fernández-Pérez, B.M.; Nagy, L.; Nagy, G.; Souto, R.M. Multi-barrel electrodes containing an internal micro-reference for the improved visualization of galvanic corrosion processes in magnesium-based materials using potentiometric scanning electrochemical microscopy. *Sens. Actuator B-Chem.* **2019**, *296*, 126625. [CrossRef]
101. Dauphin-Ducharme, P.; Asmussen, R.M.; Shoesmith, D.W.; Mauzeroll, J. In-situ Mg^{2+} release monitored during magnesium alloy corrosion. *J. Electroanal. Chem.* **2015**, *73*, 61–68. [CrossRef]
102. Salleh, S.H.; Birbilis, N.; Musameh, M.; Venkatesan, K.; Thomas, S. On the development and application of an in-house fabricated Mg^{2+} ion selective microelectrode (ISME) for assessing Mg corrosion. *J. Electrochem. Soc.* **2018**, *165*, C771–C776. [CrossRef]
103. Bastos, A.C.; Taryba, M.G.; Karavai, O.V.; Zheludkevich, M.L.; Lamaka, S.V.; Ferreira, M.G.S. Micropotentiometric mapping of local distributions of Zn^{2+} relevant to corrosion studies. *Electrochem. Commun.* **2010**, *12*, 394–397. [CrossRef]
104. Nazarov, V.A.; Taryba, M.G.; Zdrachek, E.A.; Andronchyk, K.A.; Egorov, V.V.; Lamaka, S.V. Sodium- and chloride-selective microelectrodes optimized for corrosion studies. *J. Electroanal. Chem.* **2013**, *706*, 13–24. [CrossRef]
105. Kwak, J.; Bard, A.J. Scanning electrochemical microscopy. Theory of the feedback mode. *Anal. Chem.* **1989**, *61*, 1221–1227. [CrossRef]
106. Souto, R.M.; González-García, Y.; González, S.; Burstein, G.T. Damage to paint coatings caused by electrolyte immersion as observed in situ by scanning electrochemical microscopy. *Corros. Sci.* **2004**, *46*, 2621–2628. [CrossRef]
107. Souto, R.M.; González-García, Y.; González, S. Evaluation of the corrosion performance of coil-coated steel sheet as studied by scanning electrochemical microscopy. *Corros. Sci.* **2008**, *50*, 1637–1643. [CrossRef]
108. Elkebir, Y.; Mallarino, S.; Trinh, D.; Touzain, S. Effect of physical ageing onto the water uptake in epoxy coatings. *Electrochim. Acta* **2020**, *337*, 137566. [CrossRef]
109. Duarte, R.G.; González, S.; Castela, A.S.; Ferreira, M.G.S.; Souto, R.M. Sensing polymer inhomogeneity in coated metals during the early stages of coating degradation. *Prog. Org. Coat.* **2012**, *74*, 365–370. [CrossRef]
110. Souto, R.M.; González-García, Y.; González, S. Imaging the origins of coating degradation and blistering caused by electrolyte immersion assisted by SECM. *Electroanalysis* **2009**, *21*, 2569–2574. [CrossRef]
111. Souto, R.M.; González-García, Y.; Izquierdo, J.; González, S. Examination of organic coatings on metallic substrates by scanning electrochemical microscopy in feedback mode: Revealing the early stages of coating breakdown in corrosive environments. *Corros. Sci.* **2010**, *52*, 748–753. [CrossRef]
112. Vosgien-Lacombre, C.; Bouvet, G.; Trinh, D.; Mallarino, S.; Touzain, S. Effect of pigment and temperature onto swelling and water uptake during organic coating ageing. *Prog. Org. Coat.* **2018**, *124*, 249–255. [CrossRef]
113. Trinh, D.; Vosgien-Lacombre, C.; Bouvet, G.; Mallarino, S.; Touzain, S. Use of ionic liquids in SECM experiments to distinguish effects of temperature and water organic coating swelling. *Prog. Org. Coat.* **2020**, *139*, 105438. [CrossRef]
114. Thomas, S.; Izquierdo, J.; Birbilis, N.; Souto, R.M. Possibilities and limitations of scanning electrochemical microscopy of Mg and Mg alloys. *Corrosion* **2015**, *71*, 171–183. [CrossRef]
115. Battistel, D.; Daniele, S.; Gerbasi, R.; Baldo, M.A. Characterization of metal-supported Al_2O_3 thin films by scanning electrochemical microscopy. *Thin Solid Films* **2010**, *518*, 3625–3631. [CrossRef]
116. Izquierdo, J.; Bolat, G.; Cimpoesu, N.; Trinca, L.C.; Mareci, D.; Souto, R.M. Electrochemical characterization of pulsed layer deposited hydroxyapatite-zirconia layers on Ti-21Nb-15Ta-6Zr alloy for biomedical application. *Appl. Surf. Sci.* **2016**, *385*, 368–378. [CrossRef]
117. Carbonell, D.J.; García-Casas, A.; Izquierdo, J.; Souto, R.M.; Galván, J.C.; Jiménez-Morales, A. Scanning Electrochemical Microscopy characterization of sol-gel coatings applied on AA2024-T3 substrate. *Corros. Sci.* **2016**, *111*, 625–636. [CrossRef]
118. Cornut, R.; Lefrou, C. New analytical approximation of feedback approach curves with a microdisk SECM tip and irreversible kinetic reaction at the substrate. *J. Electroanal. Chem.* **2008**, *621*, 178–184. [CrossRef]
119. Mansikkamaki, K.; Ahonen, P.; Fabricius, G.; Murtomaki, L.; Kontturi, K. Inhibitive effect of benzotriazole on copper surfaces studied by SECM. *J. Electrochem. Soc.* **2005**, *152*, B12–B16. [CrossRef]
120. Mansikkamaki, K.; Johans, C.; Kontturi, K. The Effect of Oxygen on the Inhibition of Copper Corrosion with benzotriazole. *J. Electrochem. Soc.* **2006**, *153*, B22–B24. [CrossRef]
121. Mansikkamaki, K.; Haapanen, U.; Johans, C.; Kontturi, K.; Valden, M. Adsorption of benzotriazole on the surface of copper alloys studied by SECM and XPS. *J. Electrochem. Soc.* **2006**, *153*, B311–B318. [CrossRef]

122. Izquierdo, J.; Santana, J.J.; González, S.; Souto, R.M. Uses of scanning electrochemical microscopy for the characterization of thin inhibitor films on reactive metals: The protection of copper surfaces by benzotriazole. *Electrochim. Acta* **2010**, *55*, 8791–8800. [CrossRef]
123. Izquierdo, J.; Santana, J.J.; González, S.; Souto, R.M. Scanning microelectrochemical characterization of the anti-corrosion performance of inhibitor films formed by 2-mercaptobenzimidazole on copper. *Prog. Org. Coat.* **2012**, *74*, 526–533. [CrossRef]
124. Ghilane, J.; Martin, P.; Janin, M.; Randriamahazaka, H.; Hapiot, P.; Lacroix, J. Electrochemical investigation of thin PEDOT film above an insulating substrate using scanning electrochemical microscopy. *Electrochem. Commun.* **2009**, *11*, 2304–2307. [CrossRef]
125. Herlem, G.; Kandory, A.; Tangou, R.K.; Gharbi, T.; Cattey, H.; Knorr, M.; Khatyr, A. Electrochemical deposition of a luminescent alkoxysilyl-based fluorenone film exhibiting halide sensitivity. *ECS J. Solid State Sci. Technol.* **2017**, *6*, R7–R13. [CrossRef]
126. Seegmiller, J.C.; Pereira da Silva, J.E.; Buttry, D.A.; Córdoba de Torresi, S.I.; Torresi, R.M. Mechanism of action of corrosion protection coating for AA2024-T3 based on poly(aniline)-poly(methylmethacrylate) blend. *J. Electrochem. Soc.* **2005**, *152*, B45–B53. [CrossRef]
127. Dufek, E.J.; Buttry, D.A. Characterization of Zr(IV)–Phosphonate thin films which inhibit O_2 reduction on AA2024-T3. *J. Electrochem. Soc.* **2009**, *156*, C322–C330. [CrossRef]
128. Leroux, Y.; Schaming, D.; Ruhlmann, L.; Hapiot, P. SECM investigations of immobilized porphyrins films. *Langmuir* **2010**, *26*, 14983–14989. [CrossRef] [PubMed]
129. Jiang, M.Y.; Wu, L.K.; Hu, J.M.; Zhang, J.Q. Silane-incorporated epoxy coatings on aluminum alloy (AA2024). Part 1: Improved corrosion performance. *Corros. Sci.* **2015**, *92*, 118–126. [CrossRef]
130. Li, Y.L.; Huang, Z.R.; Zhong, Q.D.; Chen, C.; Xu, Y.X. SECM applied for researching the microscopic morphology and corrosion behavior of enamel coating on mild steel in aqueous chloride solutions. *Surf. Rev. Lett.* **2018**, *25*, 1–8. [CrossRef]
131. Li, Y.; Wang, J.; Zhong, Q. A novel method for studying the corrosion resistance and microstructure of enamel coating modified on high-strength steel in 3.5 wt.% NaCl solution. *Surf. Rev. Lett.* **2019**, *27*, 1950098. [CrossRef]
132. González-García, Y.; García, S.J.; Hughes, A.E.; Mol, J.M.C. A combined redox-competition and negative-feedback SECM study of self-healing anticorrosive coatings. *Electrochem. Commun.* **2011**, *13*, 1094–1097. [CrossRef]
133. Allard, D.; Allard, S.; Brehmer, M.; Conrad, L.; Zentel, R.; Stromberg, C.; Schultze, J.W. Micro- and nanostructuring of oligo- and polythiophenes in two and three dimensions. *Electrochim. Acta* **2003**, *48*, 3137–3146. [CrossRef]
134. Calhoun, R.; Lancaster, F. Investigation of anti-corrosion coatings by scanning electrochemical microscopy (SECM). In *2012 Microsystems for Measurement and Instrumentation (MAMNA)*; IEEE: Piscataway, NJ, USA, 2012; pp. 1–3.
135. Chen, M.; Zhao, J.; Zhao, X. Scanning electrochemical microscopy studies of micropatterned copper sulfide (Cu_xS) thin films fabricated by a wet chemistry method. *Electrochim. Acta* **2011**, *56*, 5016–5021. [CrossRef]
136. Claudio-Cintrón, M.A.; Rodríguez-López, J. Scanning electrochemical microscopy with conducting polymer probes: Validation and applications. *Anal. Chim. Acta* **2019**, *1069*, 36–46. [CrossRef]
137. Fonder, G.; Volcke, C.; Csoka, B.; Delhalle, J.; Mekhalif, Z. Electrochemical and spectroscopic study of $C_{12}H_{25}X$ molecules adsorption on copper sheets, X (-SH, -S-S-, -SeH and -Se-Se-). *Electrochim. Acta* **2010**, *55*, 1557–1567. [CrossRef]
138. Guadagnini, L.; Chiavari, C.; Martini, C.; Bernardi, E.; Morselli, L.; Tonelli, D. The use of scanning electrochemical microscopy for the characterisation of patinas on copper alloys. *Electrochim. Acta* **2011**, *56*, 6598–6606. [CrossRef]
139. Jensen, M.B.; Peterson, M.J.; Jadhav, N.; Gelling, V.J. SECM investigation of corrosion inhibition by tungstate- and vanadate-doped polypyrrole/aluminum flake composite coatings on AA2024-T3. *Prog. Org. Coat.* **2014**, *77*, 2116–2122. [CrossRef]
140. Johnson, L.; Niaz, A.; Boatwright, A.; Voisey, K.T.; Walsh, D.A. Scanning electrochemical microscopy at thermal sprayed anti-corrosion coatings: Effect of thermal spraying on heterogeneous electron transfer kinetics. *J. Electroanal. Chem.* **2011**, *657*, 46–53. [CrossRef]
141. Recloux, I.; Gonzalez-Garcia, Y.; Druart, M.-E.; Khelifa, F.; Dubois, P.; Mol, J.M.C.; Olivier, M.-G. Active and passive protection of AA2024-T3 by a hybrid inhibitor doped mesoporous sol–gel and top coating system. *Surf. Coatings Technol.* **2016**, *303*, 352–361. [CrossRef]
142. Souto, R.M.; González-García, Y.; González, S. Characterization of coating systems by scanning electrochemical microscopy: Surface topology and blistering. *Prog. Org. Coat.* **2009**, *65*, 435–439. [CrossRef]
143. Walsh, D.A.; Li, L.E.; Bakare, M.S.; Voisey, K.T. Visualisation of the local electrochemical activity of thermal sprayed anti-corrosion coatings using scanning electrochemical microscopy. *Electrochim. Acta* **2009**, *54*, 4647–4654. [CrossRef]
144. Wang, W.; Xu, L.; Sun, H.; Li, X.; Zhao, S.; Zhang, W. Spatial resolution comparison of AC-SECM with SECM and their characterization of self-healing performance of hexamethylene diisocyanate trimer microcapsule coatings. *J. Mater. Chem. A* **2015**, *3*, 5599–5607. [CrossRef]
145. Xia, D.H.; Wang, J.; Wu, Z.; Qin, Z.; Xu, L.; Hu, W.; Behnamian, Y.; Luo, J.L. Sensing corrosion within an artificial defect in organic coating using SECM. *Sens. Actuator B-Chem.* **2019**, *280*, 235–242. [CrossRef]
146. Xia, Y. An in-situ degradation behavior study of MAO coating on AZ91D magnesium alloy in aqueous solutions by SECM. *Int. J. Electrochem. Sci.* **2017**, *12*, 2145–2158. [CrossRef]
147. Wittstock, G.; Asmus, T.; Wilhelm, T. Investigation of ion-bombarded conducting polymer films by scanning electrochemical microscopy (SECM). *Fresenius J. Anal. Chem.* **2000**, *367*, 346–351. [CrossRef] [PubMed]
148. Li, C.; Li, L.; Wang, C.; Zhu, Y.; Zhang, W. Study of the protection performance of self-assembled monolayers on copper with the scanning electrochemical microscope. *Corros. Sci.* **2014**, *80*, 511–516. [CrossRef]

149. Wang, J.; Wu, L.-K.; Zhou, J.-H.; Hu, J.-M.; Zhang, J.-Q.; Cao, C.-N. Construction of a novel painting system using electrodeposited SiO_2 film as the pretreatment layer. *Corros. Sci.* **2013**, *68*, 57–65. [CrossRef]
150. Alizadeh, V.; Mousavi, M.F.; Mehrgardi, M.A.; Kazemi, S.H.; Sharghi, H. Electron transfer kinetics of cytochrome c immobilized on a phenolic terminated thiol self assembled monolayer determined by scanning electrochemical microscopy. *Electrochim. Acta* **2011**, *56*, 6224–6229. [CrossRef]
151. Bard, A.J.; Denuault, G.; Lee, C.; Mandler, D.; Wipf, D.O. Scanning electrochemical microscopy: A new technique for the characterization and modification of surfaces. *Acc. Chem. Res.* **1990**, *23*, 357–363. [CrossRef]
152. Bastos, A.C.; Simões, A.M.; González, S.; González-García, Y.; Souto, R.M. Application of the scanning electrochemical microscope to the examination of organic coatings on metallic substrates. *Prog. Org. Coat.* **2005**, *53*, 177–182. [CrossRef]
153. Bouvet, G.; Trinh, D.; Mallarino, S.; Feaugas, X.; Touzain, S. In situ monitoring of organic coating swelling by dynamic mechanical analysis and scanning electrochemical microscopy. *Prog. Org. Coat.* **2016**, *96*, 13–18. [CrossRef]
154. Chen, P.-Y.; Nien, P.-C.; Wu, C.-T.; Wu, T.-H.; Lin, C.-W.; Ho, K.-C. Fabrication of a molecularly imprinted polymer sensor by self-assembling monolayer/mediator system. *Anal. Chim. Acta* **2009**, *643*, 38–44. [CrossRef]
155. Gmucová, K.; Weis, M.; Nádaždy, V.; Capek, I.; Šatka, A.; Chitu, L.; Cirák, J.; Majková, E. Effect of charged deep states in hydrogenated amorphous silicon on the behavior of iron oxides nanoparticles deposited on its surface. *Appl. Surf. Sci.* **2008**, *254*, 7008–7013. [CrossRef]
156. Hočevar, S.B.; Daniele, S.; Bragato, C.; Ogorevc, B. Reactivity at the film/solution interface of ex situ prepared bismuth film electrodes: A scanning electrochemical microscopy (SECM) and atomic force microscopy (AFM) investigation. *Electrochim. Acta* **2007**, *53*, 555–560. [CrossRef]
157. Iannucci, L.; Ríos-Rojas, J.F.; Angelini, E.; Parvis, M.; Grassini, S. Electrochemical characterization of innovative hybrid coatings for metallic artefacts. *Eur. Phys. J. Plus* **2018**, *133*, 522. [CrossRef]
158. Lhenry, S.; Leroux, Y.R.; Orain, C.; Conan, F.; Cosquer, N.; Le Poul, N.; Reinaud, O.; Le Mest, Y.; Hapiot, P. Locally induced and self-induced "electroclick" onto a self-assembled monolayer: Writing and reading with SECM under unbiased conditions. *Langmuir* **2014**, *30*, 4501–4508. [CrossRef] [PubMed]
159. Vega, J.; Scheerer, H.; Andersohn, G.; Oechsner, M. Experimental studies of the effect of Ti interlayers on the corrosion resistance of TiN PVD coatings by using electrochemical methods. *Corros. Sci.* **2018**, *133*, 240–250. [CrossRef]
160. Molina, J.; Fernández, J.; del Río, A.I.; Bonastre, J.; Cases, F. Chemical and electrochemical study of fabrics coated with reduced graphene oxide. *Appl. Surf. Sci.* **2013**, *279*, 46–54. [CrossRef]
161. Shao, Y.; Jia, C.; Meng, G.; Zhang, T.; Wang, F. The role of a zinc phosphate pigment in the corrosion of scratched epoxy-coated steel. *Corros. Sci.* **2009**, *51*, 371–379. [CrossRef]
162. Daniele, S.; Bragato, C.; Baldo, M.-A.; Battiston, G.A.; Gerbasi, R. Electrochemical behaviour of Pt-TiO_2 nanocomposites prepared by MOCVD in acidic aqueous solutions. *Mater. Sci. Forum* **2002**, *413*, 147–152. [CrossRef]
163. Yang, S.; Wang, J.; Mao, W.; Zhang, D.; Guo, Y.; Song, Y.; Wang, J.-P.; Qi, T.; Li, G.L. pH-Responsive zeolitic imidazole framework nanoparticles with high active inhibitor content for self-healing anticorrosion coatings. *Colloids Surf. A Physicochem. Eng. Asp.* **2018**, *555*, 18–26. [CrossRef]
164. Wang, X.; Liu, Q.; Chun, Y.; Li, Y.; Wang, Z. Evaluation of delamination of X80 pipeline steel coating under alternating stray current via scanning electrochemical microscopy. *J. Mater. Eng. Perform.* **2018**, *27*, 3060–3071. [CrossRef]
165. Kandory, A.; Cattey, H.; Saviot, L.; Gharbi, T.; Vigneron, J.; Frégnaux, M.; Etcheberry, A.; Herlem, G. Direct writing on copper ion doped silica films by electrogeneration of metallic microstructures. *J. Phys. Chem. C* **2017**, *121*, 1129–1139. [CrossRef]
166. Souto, R.M.; González-García, Y.; González, S. In situ monitoring of electroactive species by using the scanning electrochemical microscope. Application to the investigation of degradation processes at defective coated metals. *Corros. Sci.* **2005**, *47*, 3312–3323. [CrossRef]
167. Gabriunaite, I.; Valiūnienė, A.; Poderyte, M.; Ramanavicius, A. Silane-based self-assembled monolayer deposited on fluorine doped tin oxide as model system for pharmaceutical and biomedical analysis. *J. Pharm. Biomed. Anal.* **2020**, *177*, 112832. [CrossRef] [PubMed]
168. Pereda, M.D.; Alonso, C.; Gamero, M.; del Valle, J.A.; de Mele, M.F.L. Comparative study of fluoride conversion coatings formed on biodegradable powder metallurgy Mg: The effect of chlorides at physiological level. *Mater. Sci. Eng. C* **2011**, *31*, 858–865. [CrossRef]
169. Raj, X.J. Application of EIS and SECM studies for investigation of anticorrosion properties of epoxy coatings containing zinc oxide nanoparticles on mild steel in 3.5% NaCl solution. *J. Mater. Eng. Perform.* **2017**, *26*, 3245–3253. [CrossRef]
170. Souto, R.M.; Santana, J.J.; Fernández-Mérida, L.; González, S. Sensing electrochemical activity in polymer coated metals during the early stages of coating degradation–Effect of the polarization of the substrate. *Electrochim. Acta* **2011**, *56*, 9596–9601. [CrossRef]
171. Wan, H.; Song, D.; Li, X.; Zhang, D.; Gao, J.; Du, C. Failure mechanisms of the coating/metal interface in waterborne coatings: The effect of bonding. *Materials* **2017**, *10*, 397. [CrossRef] [PubMed]
172. Singh, A.; Lin, Y.; Liu, W.; Yu, S.; Pan, J.; Ren, C.; Kuanhai, D. Plant derived cationic dye as an effective corrosion inhibitor for 7075 aluminum alloy in 3.5% NaCl solution. *J. Ind. Eng. Chem.* **2014**, *20*, 4276–4285. [CrossRef]
173. Chlistunoff, J.; Cliffel, D.; Bard, A.J. Electrochemistry of fullerene films. *Thin Solid Films* **1995**, *257*, 166–184. [CrossRef]
174. Battistel, D.; Daniele, S.; Battaglin, G.; Baldo, M.A. A simple electrochemical strategy for the characterisation of defects in alumina-coated metal substrates. *Electrochem. Commun.* **2009**, *11*, 2195–2198. [CrossRef]

175. Battistel, D.; Daniele, S.; Fratter, D. A scanning electrochemical microscopy procedure for micropatterning Al_2O_3-thin films deposited on a platinum substrate. *Electrochim. Acta* **2012**, *78*, 557–562. [CrossRef]
176. Etienne, M.; Layoussifi, B.; Giornelli, T.; Jacquet, D. SECM-based automate equipped with a shearforce detection for the characterization of large and complex samples. *Electrochem. Commun.* **2012**, *15*, 70–73. [CrossRef]
177. Molina, J.; Fernández, J.; del Río, A.I.; Bonastre, J.; Cases, F. Chemical, electrical and electrochemical characterization of hybrid organic/inorganic polypyrrole/$PW_{12}O_{40}^{3-}$ coating deposited on polyester fabrics. *Appl. Surf. Sci.* **2011**, *257*, 10056–10064. [CrossRef]
178. Maho, A.; Kanoufi, F.; Combellas, C.; Delhalle, J.; Mekhalif, Z. Electrochemical investigation of nitinol/tantalum hybrid surfaces modified by alkylphosphonic self-assembled monolayers. *Electrochim. Acta* **2014**, *116*, 78–88. [CrossRef]
179. Molina, J.; Esteves, M.F.F.; Fernández, J.; Bonastre, J.; Cases, F. Polyaniline coated conducting fabrics. Chemical and electrochemical characterization. *Eur. Polym. J.* **2011**, *47*, 2003–2015. [CrossRef]
180. Molina, J.; Fernández, J.; Del Río, A.I.; Lapuente, R.; Bonastre, J.; Cases, F. Stability of conducting polyester/polypyrrole fabrics in different pH solutions. Chemical and electrochemical characterization. *Polym. Degrad. Stab.* **2010**, *95*, 2574–2583. [CrossRef]
181. Molina, J.; Fernández, J.; del Río, A.I.; Bonastre, J.; Cases, F. Electrochemical synthesis of polyaniline on conducting fabrics of polyester covered with polypyrrole/$PW_{12}O_{40}^{3-}$. Chemical and electrochemical characterization. *Synth. Met.* **2011**, *161*, 953–963. [CrossRef]
182. Molina, J.; Fernández, J.; Inés, J.C.; del Río, A.I.; Bonastre, J.; Cases, F. Electrochemical characterization of reduced graphene oxide-coated polyester fabrics. *Electrochim. Acta.* **2013**, *93*, 44–52. [CrossRef]
183. Oleinick, A.I.; Battistel, D.; Daniele, S.; Svir, I.; Amatore, C. Simple and clear evidence for positive feedback limitation by bipolar behavior during scanning electrochemical microscopy of unbiased conductors. *Anal. Chem.* **2011**, *83*, 4887–4893. [CrossRef]
184. Radtke, V.; Heß, C.; Souto, R.M.; Heinze, J. Electroless, electrolytic and galvanic copper deposition with the scanning electrochemical microscope. *Z. Physik. Chem.* **2006**, *220*, 393–406. [CrossRef]
185. Leonard, K.C.; Bard, A.J. The study of multireactional electrochemical interfaces via a tip generation/substrate collection mode of scanning electrochemical microscopy: The hydrogen evolution reaction for Mn in acidic solution. *J. Am. Chem. Soc.* **2013**, *135*, 15890–15896. [CrossRef]
186. Raj, X.J. Investigation into the effect of Cr_2O_3 nanoparticles on the protective properties of epoxy coatings on carbon steel in 3.5% NaCl solution by scanning electrochemical microscopy. *Prot. Met. Phys. Chem. Surf.* **2019**, *55*, 80–88.
187. Raj, X.J.; Nishimura, T. Corrosion protection performance of epoxy coated high tensile strength steel measured by scanning electrochemical microscope and electrochemical impedance spectroscopy techniques. *ISIJ Int.* **2014**, *54*, 693–699. [CrossRef]
188. Pilbáth, A.; Szabó, T.; Telegdi, J.; Nyikos, L. SECM study of steel corrosion under scratched microencapsulated epoxy resin. *Prog. Org. Coat.* **2012**, *75*, 480–485. [CrossRef]
189. Vega, J.; Scheerer, H.; Andersohn, G.; Oechsner, M. Evaluation of the open porosity of PVD coatings through electrochemical iron detection. *Surf. Coatings Technol.* **2018**, *350*, 453–461. [CrossRef]
190. Xavier, J.R. Effect of surface modified WO_3 nanoparticle on the epoxy coatings for the adhesive and anticorrosion properties of mild steel. *J. Appl. Polym. Sci.* **2020**, *137*, 48323. [CrossRef]
191. Xavier, J.R. Investigation on the effect of nano-ceria on the epoxy coatings for corrosion protection of mild steel in natural seawater. *Anti-Corros. Methods Mater.* **2018**, *65*, 38–45. [CrossRef]
192. Xavier, J.R. Investigation on the anticorrosion, adhesion and mechanical performance of epoxy nanocomposite coatings containing epoxy-silane treated nano-MoO_3 on mild steel. *J. Adhes. Sci. Technol.* **2020**, *34*, 115–134. [CrossRef]
193. Xavier, J.R.; Nishimura, T. Evaluation of the corrosion protection performance of epoxy coatings containing Mg nanoparticle on carbon steel in 0.1 M NaCl solution by SECM and EIS techniques. *J. Coat. Technol. Res.* **2017**, *14*, 395–406. [CrossRef]
194. Raj, X.J.; Nishimura, T. Evaluation of the corrosion protection performance of epoxy-coated high manganese steel by SECM and EIS techniques. *J. Fail. Anal. Prev.* **2016**, *16*, 417–426. [CrossRef]
195. Jamali, S.S.; Moulton, S.E.; Tallman, D.E.; Forsyth, M.; Weber, J.; Mirabedini, A.; Wallace, G.G. Corrosion protection afforded by praseodymium conversion film on Mg alloy AZNd in simulated biological fluid studied by scanning electrochemical microscopy. *J. Electroanal. Chem.* **2015**, *739*, 211–217. [CrossRef]
196. Jamali, S.S.; Moulton, S.E.; Tallman, D.E.; Zhao, Y.; Weber, J.; Wallace, G.G. Self-healing characteristic of praseodymium conversion coating on AZNd Mg alloy studied by scanning electrochemical microscopy. *Electrochem. Commun.* **2017**, *76*, 6–9. [CrossRef]
197. Zhang, Y.; Liu, X.; Jamali, S.S.; Hinton, B.R.W.; Moulton, S.E.; Wallace, G.G.; Forsyth, M. The effect of treatment time on the ionic liquid surface film formation: Promising surface coating for Mg alloy AZ31. *Surf. Coatings Technol.* **2016**, *296*, 192–202. [CrossRef]
198. Eckhard, K.; Chen, X.; Turcu, F.; Schuhmann, W. Redox competition mode of scanning electrochemical microscopy (RC-SECM) for visualisation of local catalytic activity. *Phys. Chem. Chem. Phys.* **2006**, *8*, 5359–5365. [CrossRef] [PubMed]
199. Souto, R.M.; Fernández-Mérida, L.; González, S. SECM imaging of interfacial processes in defective organic coatings applied on metallic substrates using oxygen as redox mediator. *Electroanalysis* **2009**, *21*, 2640–2646. [CrossRef]
200. Xiao, Y.; Gu, J.; Dai, N.; Zhang, J. Evaluation of early degradation and corrosion resistance of modified zinc coating with SiO_2 nanoparticles. *Arab. J. Sci. Eng.* **2018**, *43*, 3577–3584. [CrossRef]
201. El Jaouhari, A.; Filotás, D.; Kiss, A.; Laabd, M.; Bazzaoui, E.A.; Nagy, L.; Nagy, G.; Albourine, A.; Martins, J.I.; Wang, R.; et al. SECM investigation of electrochemically synthesized polypyrrole from aqueous medium. *J. Appl. Electrochem.* **2016**, *46*, 1199–1209. [CrossRef]

202. Gao, S.; Dong, C.; Luo, H.; Xiao, K.; Pan, X.; Li, X. Scanning electrochemical microscopy study on the electrochemical behavior of CrN film formed on 304 stainless steel by magnetron sputtering. *Electrochim. Acta* **2013**, *114*, 233–241. [CrossRef]
203. González-García, Y.; Mol, J.M.C.; Muselle, T.; De Graeve, I.; Van Assche, G.; Scheltjens, G.; Van Mele, B.; Terryn, H. SECM study of defect repair in self-healing polymer coatings on metals. *Electrochem. Commun.* **2011**, *13*, 169–173. [CrossRef]
204. Jadhav, N.; Jensen, M.B.; Gelling, V. Tungstate and vanadate-doped polypyrrole/aluminum flake composite coatings for the corrosion protection of aluminum 2024-T3. *J. Coat. Technol. Res.* **2015**, *12*, 259–276. [CrossRef]
205. Liu, X.; Hou, P.; Zhao, X.; Ma, X.; Hou, B. The polyaniline-modified TiO_2 composites in water-based epoxy coating for corrosion protection of Q235 steel. *J. Coat. Technol. Res.* **2019**, *16*, 71–80. [CrossRef]
206. Madhankumar, A.; Rajendran, N.; Nishimura, T. Influence of Si nanoparticles on the electrochemical behavior of organic coatings on carbon steel in chloride environment. *J. Coat. Technol. Res.* **2012**, *9*, 609–620. [CrossRef]
207. Qian, H.; Xu, D.; Du, C.; Zhang, D.; Li, X.; Huang, L.; Deng, L.; Tu, Y.; Mol, J.M.C.; Terryn, H.A. Dual-action smart coatings with a self-healing superhydrophobic surface and anti-corrosion properties. *J. Mater. Chem. A* **2017**, *5*, 2355–2364. [CrossRef]
208. Shi, X.; Shu, M.; Zhong, Q.; Zhang, J.; Zhou, Q.; Bui, Q.B. Investigations of local corrosion behavior of plasma-sprayed FeCr nanocomposite coating by SECM. *J. Therm. Spray Technol.* **2016**, *25*, 595–604. [CrossRef]
209. Simões, A.; Battocchi, D.; Tallman, D.; Bierwagen, G. Assessment of the corrosion protection of aluminium substrates by a Mg-rich primer: EIS, SVET and SECM study. *Prog. Org. Coat.* **2008**, *63*, 260–266. [CrossRef]
210. Simões, A.M.; Battocchi, D.; Tallman, D.E.; Bierwagen, G.P. SVET and SECM imaging of cathodic protection of aluminium by a Mg-rich coating. *Corros. Sci.* **2007**, *49*, 3838–3849. [CrossRef]
211. Wang, L.; Deng, L.; Zhang, D.; Qian, H.; Du, C.; Li, X.; Mol, J.M.C.; Terryn, H.A. Shape memory composite (SMC) self-healing coatings for corrosion protection. *Prog. Org. Coat.* **2016**, *97*, 261–268. [CrossRef]
212. Xiao, Y.; Gu, J.; Zhang, J. Semiconductor property and corrosion behavior of passive film formed on steel with zinc coating in 5% NaCl solution. *Arab. J. Sci. Eng.* **2017**, *42*, 4273–4280. [CrossRef]
213. Liu, X.; Zhang, D.; Hou, P.; Pan, J.; Zhao, X.; Hou, B. Preparation and characterization of polyelectrolyte-modified attapulgite as nanocontainers for protection of carbon steel. *J. Electrochem. Soc.* **2018**, *165*, C907–C915. [CrossRef]
214. Santana, J.J.; González-Guzmán, J.; Fernández-Mérida, L.; González, S.; Souto, R.M. Visualization of local degradation processes in coated metals by means of scanning electrochemical microscopy in the redox competition mode. *Electrochim. Acta* **2010**, *55*, 4488–4494. [CrossRef]
215. Huang, Y.; Deng, L.; Ju, P.; Huang, L.; Qian, H.; Zhang, D.; Li, X.; Terryn, H.A.; Mol, J.M.C. Triple-action self-healing protective coatings based on shape memory polymers containing dual-function microspheres. *ACS Appl. Mater. Interfaces* **2018**, *10*, 23369–23379. [CrossRef]
216. da Silva, R.M.P.; Izquierdo, J.; Milagre, M.X.; Betancor-Abreu, A.M.; de Oliveira, L.A.; Antunes, R.A.; Souto, R.M.; Costa, I. On the local corrosion behavior of coupled welded zones of the 2098-T351 Al-Cu-Li alloy produced by Friction Stir Welding (FSW): An amperometric and potentiometric microelectrochemical investigation. *Electrochim. Acta* **2021**, *373*, 137910. [CrossRef]
217. Frateur, I.; Huang, V.M.-W.; Orazem, M.E.; Pébère, N.; Tribollet, B.; Vivier, V. Local electrochemical impedance spectroscopy: Considerations about the cell geometry. *Electrochim. Acta* **2008**, *53*, 7386–7395. [CrossRef]
218. Kranz, C. Recent advancements in nanoelectrodes and nanopipettes used in combined scanning electrochemical microscopy techniques. *Analyst* **2014**, *139*, 336–352. [CrossRef] [PubMed]
219. Izquierdo, J.; Eifert, A.; Kranz, C.; Souto, R.M. In situ monitoring of pit nucleation and growth at iron passive oxide layer using combined atomic force and scanning electrochemical microscopy. *ChemElectroChem* **2015**, *2*, 1847–1856. [CrossRef]
220. Izquierdo, J.; Eifert, A.; Kranz, C.; Souto, R.M. In situ investigation of copper corrosion in acidic chloride solution using atomic force—scanning electrochemical microscopy. *Electrochim. Acta* **2017**, *247*, 588–599. [CrossRef]
221. Schulte, A.; Shuhmann, W. Constant-distance mode scanning electrochemical microscopy. In *Science, Technology and Education of Microscopy: An Overwiew*; Méndez-Vilas, A., Díaz, J., Eds.; Formatex Research Center: Badajoz, Spain, 2003; pp. 753–760.
222. Etienne, M.; Dossot, M.; Grausem, J.; Herzog, G. Combined Raman microspectrometer and shearforce regulated SECM for corrosion and self-healing analysis. *Anal. Chem.* **2014**, *86*, 11203–11210. [CrossRef]

coatings

MDPI

Article

Influence of Graphene Oxide Additions on the Corrosion Resistance of a Rust Converter Primer

Belén Díaz, Xosé Ramón Nóvoa, Carmen Pérez * and Miguel Rodríguez-Morgado

CINTECX, Universidade de Vigo, ENCOMAT Group, 36310 Vigo, Spain; belenchi@uvigo.es (B.D.); rnovoa@uvigo.es (X.R.N.); miguelromorgado@gmail.com (M.R.-M.)

* Correspondence: cperez@uvigo.es

Abstract: Graphene oxide (GO) has attractive properties, such as a two-dimensional structure. Because of its hydrophilic characteristic, well-dispersed aqueous solutions are attained. Thus, it is easily incorporated into waterborne resins. For these reasons, in the last years GO nanoparticles have been added to polymers, improving corrosion resistance. This work is focused on the performance of a commercial rust converter (RC) doped with five different RC:GO ratios, namely, 1:0; 1:0.3; 1:0.6; 1:0.9; 1:1.2 (%v/v). The X-ray diffraction technique is used to illustrate the effect of RC and RC + GO additions in the iron oxides. Zeta-potential measurements are performed to assess the surface charge of the GO particles. The corrosion resistances of the rusted samples coated with the five rust converter formulations are studied. The electrochemical impedance spectroscopy (EIS) technique and an electrical equivalent circuit are utilized to explain the experimental results. Additionally, it is found that the optimal RC:GO ratio is between 1:0.3 and 1:0.6. The better corrosion resistance reached is that of the RC:0.3GO ratio.

Keywords: rusted steel; rust converter; graphene oxide; corrosion resistance

Citation: Díaz, B.; Nóvoa, X.R.; Pérez, C.; Rodríguez-Morgado, M. Influence of Graphene Oxide Additions on the Corrosion Resistance of a Rust Converter Primer. *Coatings* **2022**, *12*, 345. https://doi.org/10.3390/coatings12030345

Academic Editors: Fernando Pedraza and Ludmila B. Boinovich

Received: 9 February 2022
Accepted: 4 March 2022
Published: 6 March 2022

1. Introduction

The use of paints for the protection of steel structures is one of the most common strategies to mitigate corrosion. The success of this procedure is strongly dependent on the surface preparation of the metal. The optimum condition implies complete removal of rust; however, this is not always possible due to diverse reasons such as the inaccessibility of the elements, the repairing of large structures where conventional rust cleaning methods are expensive, or environmental constraints [1]. In these situations, an attractive alternative is the stabilization of the oxide layer before painting by the use of rust converters, which essentially consist of chemical formulations able to transform the iron oxides into a more compact and adherent layer. These compounds are designed to be applied on corroded surfaces [2,3]. The commercial rust converter (RC) formulations include two main components: the active one that reacts with the rust to generate new less reactive compounds, and the polymeric component responsible for the continuous film formation [3]. Most common formulations are based on tannic or phosphoric acids [4–8], and more recently they are also based on gallic acid [9–11] as an active component. Although they represent environmentally good alternatives to contaminant pretreatments, their efficiency is subject to controversy. Several parameters condition the final result, for example, the nature of the rust to be converted (composition, age, or thickness) that is determined by the type of atmosphere (presence of aggressive electrolytes, pH, etc.) [8,12,13], and the contact time between the RC and the rust [14]. This latter aspect is related to the chemical nature of the active component in the RC. There is a general agreement that the stabilization of the rust is based on the ability of these compounds to react with the iron oxides/hydroxides creating a passive layer [15] that can be accompanied by a sealing effect [16]. However, this protection is limited or even not present, since the interaction between the oxides and the rust converter solutions affects only the outermost rust surface [4,17,18]. Ocampo et al.

suggest that their effect can be even harmful in the case of high chloride contamination [1]. Thus, developing new RC formulations with good long-term protection is a challenging task. To our best knowledge, few studies deal with this aspect [10,19].

The incorporation of corrosion inhibitors to the commercial RC may be an alternative strategy, following the same route as employed in traditional paint formulations. In a previous work [11], we demonstrated the benefit provided by the addition of oxide (GO) to a commercial rust converter. The choice of GO as a pigment for anticorrosive coatings was based on some advantageous features that it possesses. It is a two-dimensional nanomaterial; i.e., it has a very high aspect ratio, thus it would provide a good barrier effect. Besides, it has a hydrophilic character due to the oxygen-containing functional groups (epoxy and hydroxyl at the planar surface, and carbonyl, carboxyl, ester, ether, phenol, among others, at the edges). This characteristic encourages the stabilization of the aqueous GO dispersions and also provides active points to be covalently bonded to other species [20,21]. Functionalization of the GO is a common approach used to improve the compatibility of the GO with the polymeric matrix, thus avoiding dispersion or agglomeration problems [22,23]. However, the experimental procedure usually is complex, which restricts its implementation on an industrial scale. In our proposal, the GO was added as an aqueous dispersion over commercial water-based RC, which simplifies the mixing process and the dispersion of the mixture was good [11]. Using a similar procedure, the present work is focused on the optimization of the GO concentration that provides the best anticorrosive properties of the commercial rust converter.

2. Materials and Methods

Mild steel samples with 80 mm × 30 mm × 1 mm dimensions and nominal chemical composition (wt.%): C (0.096), Mn (0.350), Si (0.183), Cr (0.123), Ni (0.161), Mo (0.06), P (0.022), S (0.033), Co (0.046), Sn (0.041), and Fe (balance), were degreased using soapy water and afterward rinsed with deionized water and dried. The clean steel plates were prerusted in a humidity chamber at 100% relative humidity and room temperature until the generation of a uniform rust layer; this process takes about two weeks. The corroded samples were brushed to remove the loose rust before applying the RC formulations. The detached oxides were collected and dried in a furnace at about 40 °C. They were then milled to obtain a fine powder and dry-stored to perform further experiments.

The commercial rust converter named Oroprimer® 440 (produced by Oropal, Irurena Group, Azpeitia, Spain) was used. It is a water emulsion based on vinyl chloride–acrylic ester copolymer, with gallic acid as the active component; the density is 1.2 g. mL^{-1} and the total solid content is 50%. The graphene oxide employed is a GO water dispersion with 0.4 wt.% concentration from Graphenea® (Graphenea Inc., Boston, MA, USA). Five RC:GO ratios, 1:0; 1:0.3; 1:0.6; 1:0.9; and 1:1.2 (%v/v), were studied. The mixtures were then sonicated for 2 min to obtain a homogenous suspension that was applied by brush on the rusted steel samples and cured for 48 h at room temperature. Figure 1 illustrates the aspect of the rusted steel samples coated with the RC/GO formulations. As can be appreciated, a thin and uniform film was obtained, the average thickness was 42 ± 2.6 µm.

The surface charge of the graphene oxide particles when they are in contact with the gallic acid was determined by zeta potential measurements. For that, gallic aqueous solutions (15 $mg \cdot mL^{-1}$) with different concentrations of GO, 0.28 $mg \cdot mL^{-1}$, 0.58 $mg \cdot mL^{-1}$, 0.74 $mg \cdot mL^{-1}$ and 0.92 $mg \cdot mL^{-1}$, were prepared. These concentrations were determined taking into account that the amount of gallic acid in the rust converter is around 10% (information provided by the manufacturer) and the RC/GO ratios of 1:0.3; 1:0.6; 1:0.9 and 1:1.2 proposed in this study. The measurements were performed using a Malvern Zetasizer NanoSizer NanoZS® (Malvern Panalytical, Worcestershire, UK) instrument. The reproducibility was verified by performing a minimum of 10 measurements per set. The zeta potential was calculated from the electrophoretic mobility using the Smoluchowski model, wherein the thickness of the electrical double layer is assumed to be small in comparison with the particle size.

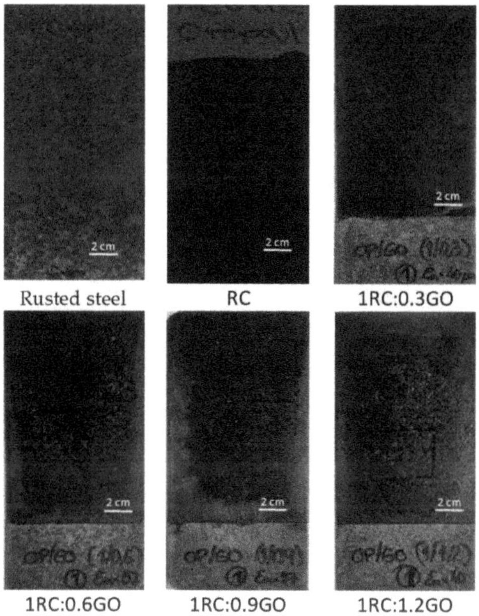

Figure 1. Pictures of rusted steel samples that were coated using the different RC:GO ratios, with 80 mm × 30 mm × 1 mm dimensions.

The interaction between the rust converter and iron oxides was assessed by the X-ray diffraction (XRD) technique. Samples were prepared by mixing the rust powder with the RC/GO formulations. After drying, these were again milled to obtain a fine powder. The equipment employed was an X-ray diffractometer PANalytical X'Pert® (Malvern Panalytical, Worcestershire, UK) powder with the monochromatic CuK_α radiation ($\lambda = 1.5$ Å) from $2\theta = 15°$ to $2\theta = 70°$.

The protective features of the RC/GO formulations applied on the rusted steels were assessed using the electrochemical impedance spectroscopy (EIS) technique. The electrochemical cell was a three-electrode arrangement, wherein the working electrode was the coated steel defining a working area of $S = 1$ cm^2, the counter electrode was a large graphite sheet (28 cm^2), and a saturated calomel electrode (SCE) was used as the reference electrode. The electrolyte was 0.06 M NaCl solution confined in a volume of about 80 mL. The reason to choose this low chloride concentration is based on the characteristics of the rust converter; it is designed as a primer coat, thus, it has poor barrier properties. The use of a low aggressive electrolyte allows tracking the behavior of the different formulations at long immersion times. The equipment was an Autolab 30 potentiostat from Ecochemie® (Metrohm AG, Herisau, Switzerland), at null DC current and applying a 10 mV_{rms} sinusoidal amplitude. The frequency ranged from 100 kHz to 10 mHz, with 7 points per decade. At least three samples were tested for each system to verify the reproducibility.

3. Results and Discussion

3.1. *Surface Charge of the Graphene Oxide Particles*

The zeta-potential measurements provide useful information about the stability of the GO suspensions in the rust converter. For comparative purposes, zeta-potential measurements of GO aqueous solutions without acid gallic were also performed. The results are depicted in Figure 2.

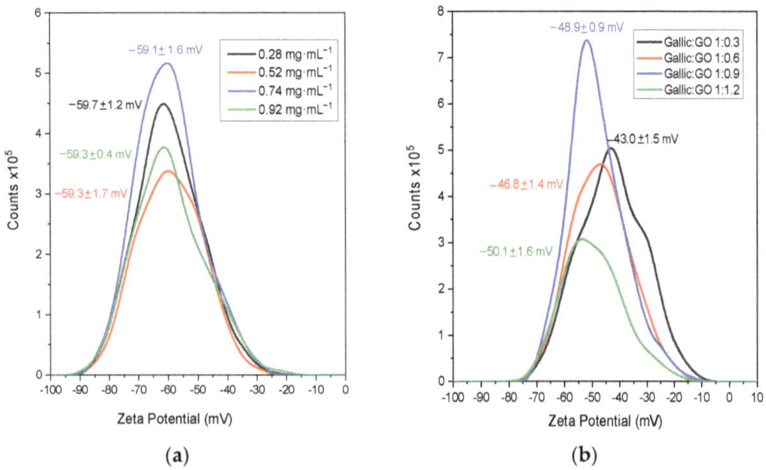

Figure 2. Zeta-potential distributions of GO in (**a**) aqueous solutions and (**b**) aqueous gallic acid solutions. The concentrations correspond to the amount of GO in the RC/GO formulations studied.

Figure 2a shows the zeta-potential values of the GO aqueous suspensions at different concentrations. The low value obtained, −59 mV, indicate very stable suspensions, regardless of the concentration. It is widely accepted that absolute values higher than 30 mV lead to stable suspensions due to interparticle electrostatic repulsion [24]. On the other hand, the negative zeta potential values are owing to the electronegative functional groups present in the GO particles [25]. The effect of the gallic acid is shown in Figure 2b. The zeta-potential values remain well below −30 mV, which confirms the stability of these suspensions. However, the recorded values are less negative than those in aqueous solutions. The reason should be twofold: on one hand, the parent solution is a concentrated gallic acid solution (15 mg·mL^{-1}) with high ionic strength, which reduces the thickness of the double layer and the repulsive forces, so that increases the zeta potential [26]. On the other hand, the partial protonation of the functional groups of the GO particles neutralizes the surface negative charge. This protonation seems more effective in dilute solutions, where the amount of GO particles is lower. This fact can explain the tendency for zeta potential to decrease when the GO concentration increases.

3.2. X-ray Characterization

The interaction between the graphene oxide particles and the commercial rust converter was also verified by XRD. Figure 3 shows a detail of the X-ray diffractograms corresponding to the lowest and highest RC:GO ratios tested, in which the more characteristic diffraction peaks of gallic acid are depicted. The intensity of those peaks is markedly lower when the amount of GO is higher, which suggests certain interaction between both species. This result is not unexpected considering the reducing character of the gallic acid and is in agreement with the work of Li et al., in which the GO is reduced by gallic acid; this acts not only as a reductant agent but also as a stabilizer of the reduced graphene oxide (rGO) suspensions that were obtained [27]. The X-ray pattern of pure graphene oxide is also included, with a characteristic peak centered at $2\theta = 10°$. As can be seen, this peak is not observed in any of the RC/GO mixtures, owing to the low GO amount in the formulations. It also corroborates the interaction between gallic acid and GO. Similar results were obtained using other polyphenols such as tannic acid [28].

The XRD technique was also used to characterize the iron oxides/hydroxides generated and several mixtures of rust with the RC. The diffractograms obtained for the rust powder and its combination with the RC are shown in Figure 4. The main crystalline compounds identified were lepidocrocite γ-FeO(OH) (ref. code 00-008-0098) and magnetite Fe_3O_4 (ref. code 00-019-0629); traces of goethite, α-FeO(OH) (ref. code 00-029-0713) were also detected. The same species were found by other researchers, even though the relative amounts may differ depending on the conditions in which the rust was generated [16,29,30]. The main crystalline compounds were also clearly detected when the rust was in contact with the RC, although there was a significant decrease in the intensity of the peaks. This expected result confirms the interaction between the rust converter and the iron oxides/hydroxides, and is in good agreement with other researchers [1,13].

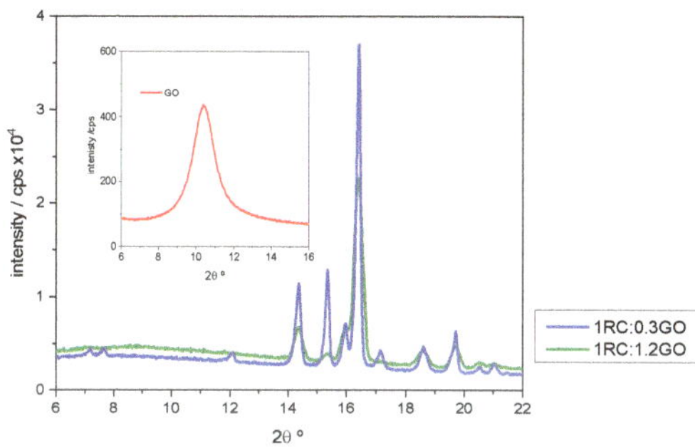

Figure 3. Detail of the X-ray patterns corresponding to 1RC:0.3GO and 1RC:1.2GO ratios. The X-ray pattern of GO is given in the insert.

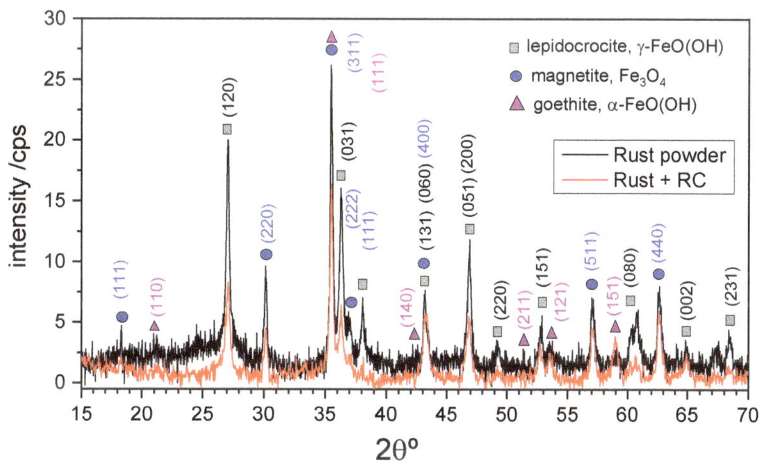

Figure 4. X-ray patterns obtained for dried rust powder and rust + RC mixture after being in contact for 15 days. The identification of the main crystalline compounds is also included together with the diffraction planes in each identified species based on refs. [31–33].

When the rust is mixed with the RC/GO formulations the intensity of the peaks also decreases, as Figure 5 illustrates. Although there is no clear tendency, the reduction is broadly less pronounced in the RC + GO blends. Even though it is not totally understood, this behavior may be related to the previous interaction between the gallic acid and the GO; thus, a smaller amount of the active component of the RC (i.e., gallic acid) would be available to react with the rust. As expected, the GO peak is not observed, as was already evident in the RC/GO mixtures.

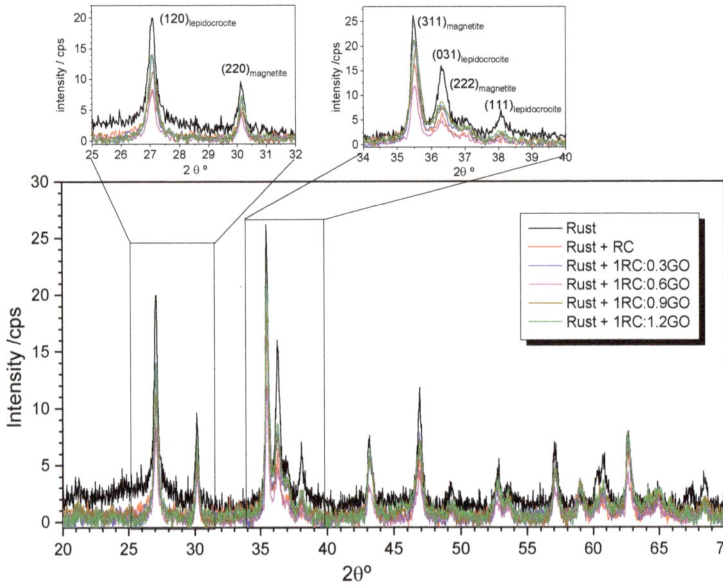

Figure 5. X-ray diffractograms obtained for rust powder and the different RC/GO mixtures tested, after being in contact for 15 days.

3.3. Electrochemical Study of Rusted Steel Coated with the RC/GO Treatments

EIS measurements were performed to assess the corrosion resistance of the RC/GO formulations, and some examples are shown in Figure 6.

As can be seen, the better behavior corresponds to 1RC:0.3GO ratio not only initially but also for long immersion periods. It is noticeable that the well-differentiated behavior observed depends on the RC/GO ratio at short immersion periods. This suggests that the larger amount of GO involves the worst corrosion properties of the films, perhaps because of certain agglomeration of the particles. In addition, the interaction with gallic acid leads to a larger amount of rGO when the RC/GO ratio is higher, which can accelerate the cathodic reaction and corrosion rate. Chaudhry et al. reported the harmful effect of the rGO addition to self-crosslinked polyvinyl butyral (PVB) polymer [34]. Generally, the impedance of all the RC/GO formulations tends to decrease with immersion time, except that of 1RC:0.3GO, which increases at longer exposure periods. A more detailed impedance evolution of each formulation is shown in Figure 7.

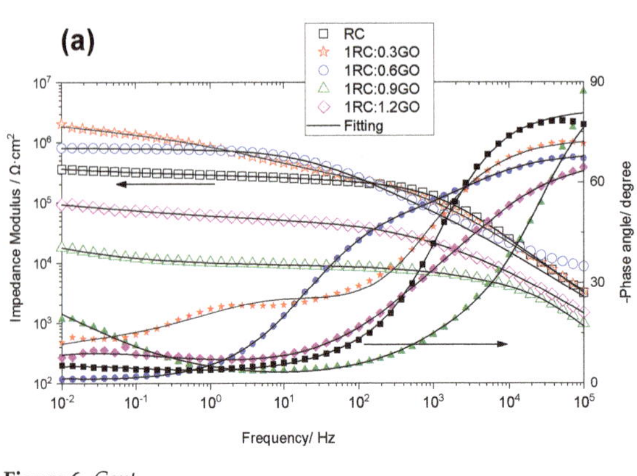

Figure 6. *Cont.*

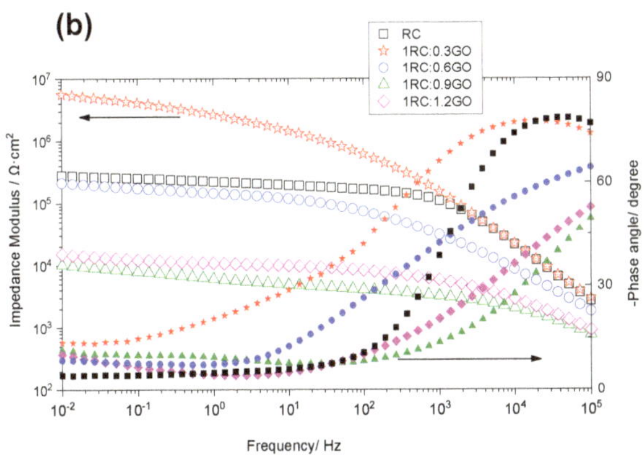

Figure 6. Impedance modulus (open symbols) and phase angle (filled symbols) of the rusted steel coated with the RC/GO formulations after (**a**) 4 days and (**b**) 21 days of immersion in 0.06 M NaCl solution.

As aforementioned, only 1RC:0.3GO formulation provides lasting corrosion resistance, even acquiring higher impedance values at the end of the immersion period. The complexity of the Nyquist plots suggests the emergence of several processes. The electrical equivalent circuit (EEC) used to model the observed behavior is shown in Figure 7f. In this, R_e denotes the electrolyte resistance. The high-frequency time constant, R_fC_f, is related to the barrier properties of the conversion coating. The time constant located at the medium frequency range, $R_{ct}C_{dl}$, accounts for the charge transfer resistance and double layer capacitance of the corrosion process. Both time constants are affected by Cole–Cole type dispersion (α_f and α_{dl}) [35]. The low-frequency time constant, Z_d, can be attributed to the oxygen diffusion toward the metallic surface. This physical interpretation is previously given in other works [36]. The total impedance is given in Equation (1):

$$Z(\omega) = R_e + \frac{R_f}{(j\omega R_fC_f)^{\alpha_f} + \frac{1}{1+Z_2/R_f}}, \quad (1)$$

The $Z_2(\omega)$ impedance is given by Equation (2):

$$Z_2(\omega) = \frac{R_{ct}}{(j\omega R_{ct}C_{dl})^{\alpha_{dl}} + \frac{1}{1+Z_d/R_{ct}}}, \quad (2)$$

Additionally, $Z_d(\omega)$ is:

$$Z_d(\omega) = R_d\frac{\tanh\sqrt{(j\omega\tau_d)}}{\sqrt{(j\omega\tau_d)}}, \quad (3)$$

In Equation (3), $Z_d(\omega)$ corresponds to semi-infinite diffusion process with a characteristic time constant $\tau_d = \delta^2/D$, where δ is the thickness of the Nernst diffusion layer, defined by the concentration gradient established by the oxygen consumption at the metal/film interface, and D accounts for the diffusion coefficient of oxygen [37]. The good agreement between the fitting and experimental data can be seen in Figures 6a and 7a–e.

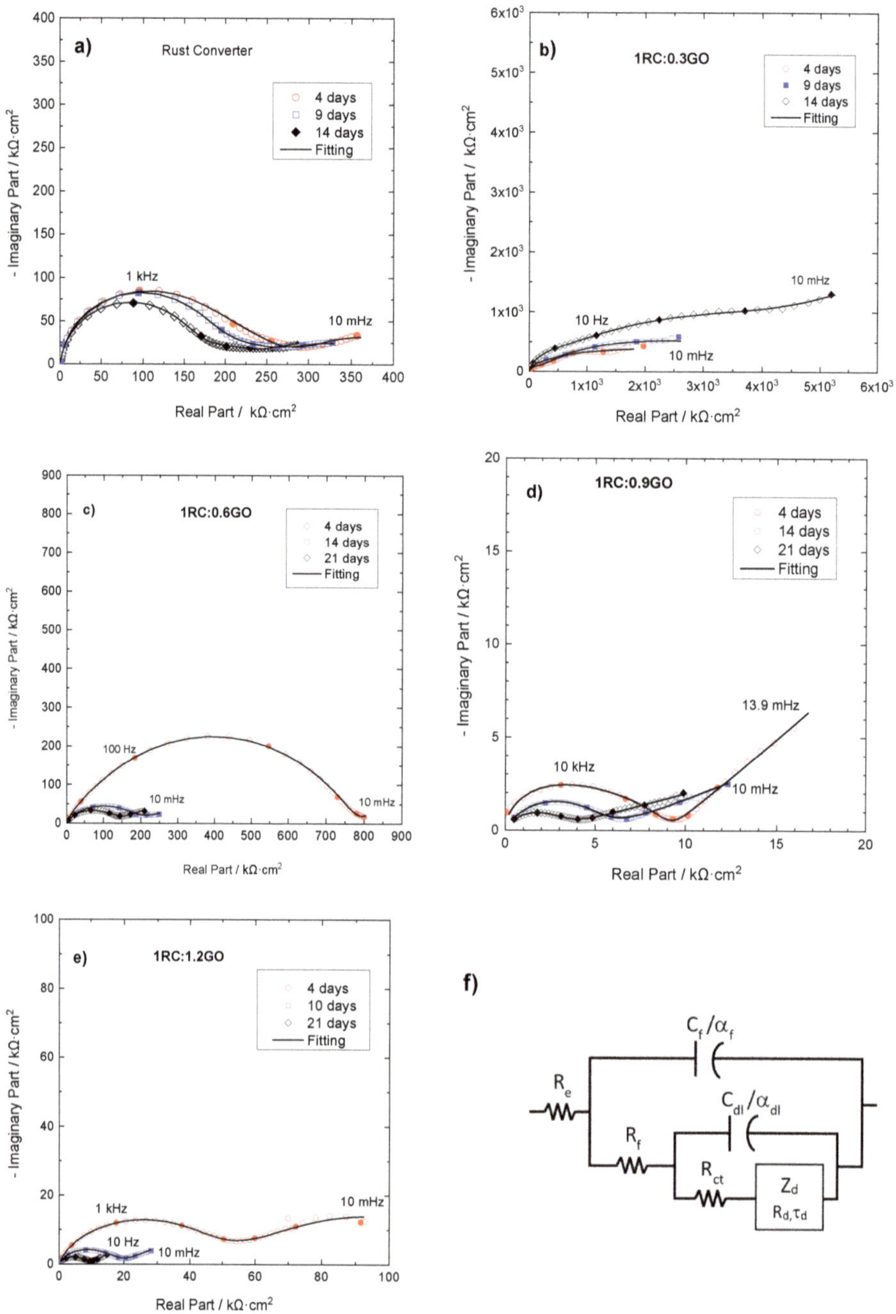

Figure 7. (**a**–**e**) Nyquist plots showing the evolution of the different RC/GO formulations with immersion time. The best-fit curve obtained for each measurement is also shown. (**f**) Electrical equivalent circuit (EEC) used for modeling the data. The meaning of the different elements is described in the text.

Figure 8a,b shows the evolution of the high-frequency time constant, R_fC_f, as aforementioned is related to the barrier properties of the conversion coating. The higher resistance values correspond to 1RC:0.3GO, followed by the undoped RC system. Initially, the 1RC:0.6GO formulation also exhibits high resistance values; however, these rapidly decrease with immersion time. On the other side, the 1RC:0.9GO and 1RC:1.2GO ratios have about one order of magnitude lower resistance values. These results illustrate the benefit afforded by small additions of GO; it seems that the conversion coatings can be effectively formed and, additionally, the GO improves the barrier effect owing to its 2D structure. As the amount of GO increases, less gallic acid (the reactive component of the RC) is available to react with the iron oxides/hydroxides, thus generating a more defective conversion coating. Parallel to the resistance evolution, the film capacitance increase observed for 1RC:0.6GO, 1RC:0.9GO, and 1RC:1.2GO formulations indicates the higher water uptake in the coating. The steady capacitance values exhibited by the other ratios corroborate the generation of a more compact layer.

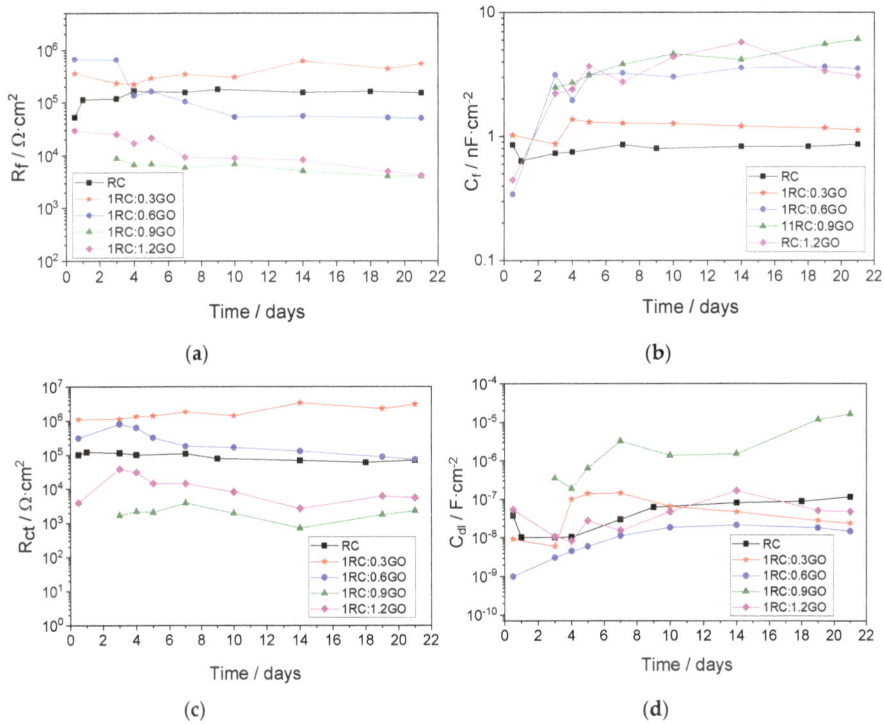

Figure 8. Evolution of the fitting parameters corresponding to the high-frequency time constant (**a**) R_f and (**b**) C_f, and to the middle frequency time constant (**c**) R_{ct} and (**d**) C_{dl}.

The evolution of the time constant associated with the corrosion process, $R_{ct}C_{dl}$, is shown in Figure 8c,d. The higher charge transfer means the lower corrosion rate [38], which corresponds to the 1RC:0.3GO system. As expected, higher corrosion rates are observed in 1RC:0.9GO and 1RC:1.2GO formulations. The double layer capacitance can be correlated to the active area, A_d, according to Equation (4) [39]:

$$A_d = \frac{C_{dl}}{C_{dl}^{o}}, \quad (4)$$

where C^o_{dl} is the specific double layer capacitance of the bare metal, which is considered constant over the immersion time. Accordingly, C_{dl} is a measure of the active area that, as can be seen in Figure 8d, increases with immersion time in all samples. However, performing an estimation of A_d is not easy for rusted samples, since the value of C^o_{dl} depends on the surface roughness. In the present study, 160 $\mu F \cdot cm^{-2}$ was adopted, which is characteristic of a very rough metallic surface [40]. The active area resulted to be less than 0.1% for all systems except 1RC:0.9GO with a larger active area (about 10%), in agreement with a lower charge transfer resistance.

It is worth commenting about the R_{ct} and C_{dl} values for the 1RC:0.6GO system. Although the conversion layer seems to have lower barrier properties than the undoped RC, the corrosion rate and the active surface are lower, perhaps because of the electrolyte uptakes to the film, although it does not reach the metal surface.

As aforementioned, the low-frequency time constant accounts for the oxygen diffusion transport from the bulk solution, through the layer pores, to the cathodic sites. The evolution of the associated resistance is depicted in Figure 9a. As can be seen, its evolution follows approximately the pattern observed for charge transfer resistance, the higher values correspond to the 1RC:0.3GO formulation and the lowest to the 1RC:9.9GO and 1RC:1.2GO ratios. This tendency corroborates the benefit provided by small GO additions to the RC, mainly the 1RC:0.3GO ratio.

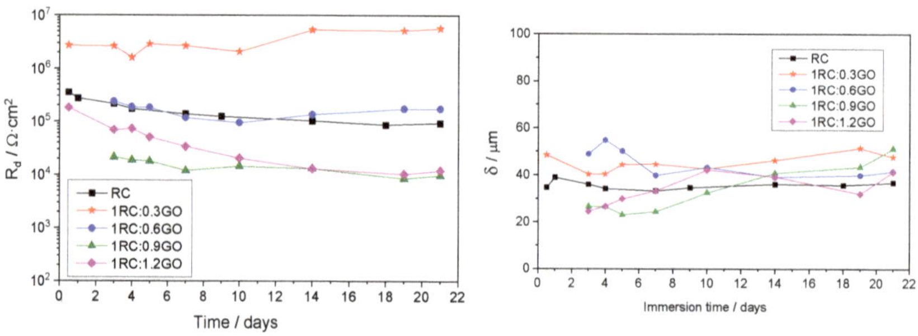

(a) (b)

Figure 9. Evolution with the immersion time of (**a**) R_d and (**b**) the diffusion length δ.

The diffusion time constant, τ_d, relates the diffusion coefficient, D, with the oxygen diffusion length, δ, by the expression $\tau_d = \delta^2/D$, previously mentioned. Assuming a typical value for oxygen diffusion through a polymer about 10^{-7} $cm^2 \cdot s^{-1}$ [41], estimation of δ can be performed. The results are shown in Figure 9. As can be seen, roughly the diffusion length corresponds to the conversion layer thickness, although certain fluctuation is observed, indicative of the dynamic nature of the process.

4. Conclusions

Different amounts of GO were added to a commercial RC in order to assess the optimum formulation range. Based on the results obtained, the following conclusions can be drawn:

The interaction between the gallic acid and the GO led to a decrease in the surface charge of the GO particles, even though the zeta-potential values confirm the stability of the colloidal suspensions for all the RC:GO ratios. This interaction was also corroborated by XRD, where the intensity of the peaks corresponding to the gallic acid markedly decreased at higher RC:GO ratios.

Additionally, the XRD technique was used to identify the main crystalline oxide/ hydroxide compounds obtained from the rusted steel. They were lepidocrocite γ-FeO(OH) and magnetite Fe_3O_4. The intensity of the peaks for both species strongly decreased when they were in contact with the RC.

The comparative study of the protection properties of all formulations applied on rusted steel was performed by EIS. Generally, the impedance decreased with immersion time; the only exception was the 1RC:0.3GO ratio, in which impedance values increased with time. On the other hand, lower impedance values were obtained for 1RC:0.9GO and 1RC:1.2GO ratios.

The electrical equivalent circuit used to model the observed behavior included three time constants. The high-frequency time constant was correlated to the dielectric properties of the conversion layer, the middle-frequency time constant accounted for the corrosion process, and the low-frequency time constant was associated with the oxygen diffusion through the layer pores.

Better behavior was observed for 1RC:0.3GO ratio, followed by the 1RC:0.6GO formulation. The benefit provided by the addition of small amounts of GO can be explained by taking into account its 2D morphology that increases the barrier properties. Additionally, the gallic acid may partially reduce the GO (rGO), which has hydrophobic character. If the GO amount is too high, agglomeration problems can arise, generating a more porous layer. On the other hand, the presence of higher amounts of rGO provides a major cathodic surface, which can accelerate the corrosion process.

Author Contributions: Conceptualization, C.P. and X.R.N.; methodology, B.D. and M.R.-M.; formal analysis, B.D. and X.R.N.; investigation, C.P. and M.R.-M.; resources, B.D. and M.R.-M.; writing—original draft preparation, C.P. and X.R.N.; writing—review and editing, C.P.; supervision, X.R.N. All authors have read and agreed to the published version of the manuscript.

Funding: This research received no external funding.

Institutional Review Board Statement: Not applicable.

Informed Consent Statement: Not applicable.

Data Availability Statement: All the data generated during this study are included in this article.

Conflicts of Interest: The authors declare no conflict of interest.

References

1. Ocampo, L.M.; Margarit, I.C.P.; Mattos, O.R.; Córdoba-de-Torresi, S.I.; Fragata, F.L. Performance of rust converter based in phosphoric and tannic acids. *Corros. Sci.* **2004**, *46*, 1515–1525. [CrossRef]
2. Feliu, S.; Galván, J.C.; Feliu, S.; Bastidas, J.M.; Simancas, J.; Morcillo, M.; Almeida, E.M. An electrochemical impedance study of the behaviour of some pretreatments applied to rusted steel surfaces. *Corros. Sci.* **1993**, *35*, 1351–1358. [CrossRef]
3. Saji, V.S. Progress in rust converters. *Prog. Org. Coat.* **2019**, *127*, 88–99. [CrossRef]
4. Morcillo, M.; Feliu, S.; Simancas, J.; Bastidas, J.M.; Galvan, J.C.; Feliu, S.; Almeida, E.M. Corrosion of Rusted Steel in Aqueous Solutions of Tannic Acid. *Corrosion* **1992**, *48*, 1032–1039. [CrossRef]
5. Nasrazadani, S. The application of infrared spectroscopy to a study of phosphoric and tannic acids interactions with magnetite (Fe_3O_4), goethite (α-FEOOH) and lepidocrocite (γ-FeOOH). *Corros. Sci.* **1997**, *39*, 1845–1859. [CrossRef]
6. Qian, B.; Hou, B.; Zheng, M. The inhibition effect of tannic acid on mild steel corrosion in seawater wet/dry cyclic conditions. *Corros. Sci.* **2013**, *72*, 1–9. [CrossRef]
7. Gust, J. Application of infrared spectroscopy for investigation of rust phase component conversion by agents containing oak tannin and phosphoric acid Indexed keywords. *Corrosion.* **1991**, *47*, 453–457. [CrossRef]
8. Rahim, A.A.; Rocca, E.; Steinmetz, J.; Jain Kassim, M. Inhibitive action of mangrove tannins and phosphoric acid on pre-rusted steel via electrochemical methods. *Corros. Sci.* **2008**, *50*, 1546–1550. [CrossRef]
9. Jia, Y.; Ren, N.; Yue, H.; Deng, J.; Liu, Y. Preparation and properties of natural gallic acid based rust conversion emulsion. *Pigment Resin Technol.* **2016**, *45*, 191–198. [CrossRef]
10. Wang, X.; Zhu, Q.; Liu, X.; Hou, B. Rust Conversion Performance of Phosphoric Acid-Gallic Acid in Vinyl Chloride Acrylic Emulsion. *Coatings* **2021**, *11*, 152. [CrossRef]
11. Díaz, B.; Figueroa, R.; Nóvoa, X.R.; Pérez, C.; Pintos, A. The corrosion protection afforded by a commercial rust converter doped with graphene oxide. *Electrochim. Acta* **2020**, *342*, 136096. [CrossRef]
12. Xu, W.; Han, E.H.; Wang, Z. Effect of tannic acid on corrosion behavior of carbon steel in NaCl solution. *J. Mater. Sci. Technol.* **2019**, *35*, 64–75. [CrossRef]

13. Collazo, A.; Nóvoa, X.R.R.; Pérez, C.; Puga, B. EIS study of the rust converter effectiveness under different conditions. *Electrochim. Acta* **2008**, *53*, 7565–7574. [CrossRef]
14. Barrero, C.A.; Ocampo, L.M.; Arroyave, C.E. Possible improvements in the action of some rust converters. *Corros. Sci.* **2001**, *43*, 1003–1018. [CrossRef]
15. Favre, M.; Landolt, D. The influence of gallic acid on the reduction of rust on painted steel surfaces. *Corros. Sci.* **1993**, *34*, 1481–1494. [CrossRef]
16. Gust, J.; Bobrowicz, J. Sealing and Anti-Corrosive Action of Tannin Rust Converters. *Corrosion* **1993**, *49*, 24–30. [CrossRef]
17. Favre, M.; Landolt, D.; Hoffman, K.; Stratmann, M. Influence of gallic acid on the phase transformation in iron oxide layers below organic coatings studied with Moessbauer spectroscopy. *Corros. Sci.* **1998**, *40*, 793–803. [CrossRef]
18. Galván, J.C.; Simancas, J.; Morcillo, M.; Bastidas, J.M.; Almeida, E.; Feliu, S. Effect of treatment with tannic, gallic and phosphoric acids on the electrochemical behaviour of rusted steel. *Electrochim. Acta* **1992**, *37*, 1983–1985. [CrossRef]
19. Li, J.; Ge, S.; Wang, J.; Du, H.; Song, K.; Fei, Z.; Shao, Q.; Guo, Z. Water-based rust converter and its polymer composites for surface anticorrosion. *Colloids Surfaces A Physicochem. Eng. Asp.* **2018**, *537*, 334–342. [CrossRef]
20. Xue, B.; Yu, M.; Liu, J.; Li, S.; Xiong, L.; Kong, X. Corrosion Protective Properties of Silane Functionalized Graphene Oxide Film on AA2024-T3 Aluminum Alloy. *J. Electrochem. Soc.* **2016**, *163*, C798–C806. [CrossRef]
21. Ramezanzadeh, B.; Niroumandrad, S.; Ahmadi, A.; Mahdavian, M.; Mohamadzadeh Moghadam, M.H. Enhancement of barrier and corrosion protection performance of an epoxy coating through wet transfer of amino functionalized graphene oxide. *Corros. Sci.* **2016**, *103*, 283–304. [CrossRef]
22. Yu, Z.; Di, H.; Ma, Y.; He, Y.; Liang, L.; Lv, L.; Ran, X.; Pan, Y.; Luo, Z. Preparation of graphene oxide modified by titanium dioxide to enhance the anti-corrosion performance of epoxy coatings. *Surf. Coat. Technol.* **2015**, *276*, 471–478. [CrossRef]
23. Cui, J.; Xiong, Z.; Qiu, H.; LI, J.; Yang, J. Functionalized graphene oxide: Carrier for corrosion inhibitor and barrier in waterborne epoxy coatings. *Compos. Part A Appl. Sci. Manuf.* **2021**, *144*, 106354. [CrossRef]
24. Li, M.J.; Liu, C.M.; Xie, Y.B.; Cao, H.B.; Zhao, H.; Zhang, Y. The evolution of surface charge on graphene oxide during the reduction and its application in electroanalysis. *Carbon N. Y.* **2014**, *66*, 302–311. [CrossRef]
25. Hu, X.; Yu, Y.; Hou, W.; Zhou, J.; Song, L. Effects of particle size and pH value on the hydrophilicity of graphene oxide. *Appl. Surf. Sci.* **2013**, *273*, 118–121. [CrossRef]
26. Lunardi, C.N.; Gomes, A.J.; Rocha, F.S.; De Tommaso, J.; Patience, G.S. Experimental methods in chemical engineering: Zeta potential. *Can. J. Chem. Eng.* **2021**, *99*, 627–639. [CrossRef]
27. Li, J.; Xiao, G.; Chen, C.; Li, R.; Yan, D. Superior dispersions of reduced graphene oxide synthesized by using gallic acid as a reductant and stabilizer. *J. Mater. Chem. A* **2013**, *1*, 1481–1487. [CrossRef]
28. Lei, Y.; Tang, Z.; Liao, R.; Guo, B. Hydrolysable tannin as environmentally friendly reducer and stabilizer for graphene oxide. *Green Chem.* **2011**, *13*, 1655–1658. [CrossRef]
29. Ross, T.K.; Francis, R.A. The treatment of rusted steel with mimosa tannin. *Corros. Sci.* **1978**, *18*, 351–361. [CrossRef]
30. de la Fuente, D.; Díaz, I.; Simancas, J.; Chico, B.; Morcillo, M. Long-term atmospheric corrosion of mild steel. *Corros. Sci.* **2011**, *53*, 604–617. [CrossRef]
31. Chen, Y.H.; Li, F.A. Kinetic study on removal of copper(II) using goethite and hematite nano-photocatalysts. *J. Colloid Interface Sci.* **2010**, *347*, 277–281. [CrossRef] [PubMed]
32. Compeán-Jasso, M.E.; Ruiz, F.; Martínez, J.R.; Herrera-Gómez, A. Magnetic properties of magnetite nanoparticles synthesized by forced hydrolysis. *Mater. Lett.* **2008**, *62*, 4248–4250. [CrossRef]
33. Antony, H.; Legrand, L.; Maréchal, L.; Perrin, S.; Dillmann, P.; Chaussé, A. Study of lepidocrocite γ-FeOOH electrochemical reduction in neutral and slightly alkaline solutions at 25 °C. *Electrochim. Acta* **2005**, *51*, 745–753. [CrossRef]
34. Chaudhry, A.U.; Mittal, V.; Mishra, B. Inhibition and promotion of electrochemical reactions by graphene in organic coatings. *RSC Adv.* **2015**, *5*, 80365–80368. [CrossRef]
35. MacDonald, J.R. *Impedance Spectroscopy: Emphasizing Solid Materials and Systems*; John Wiley & Sons, Inc.: New York, NY, USA, 1987; ISBN 0-471-83122-0.
36. Pérez, C.; Collazo, A.; Izquierdo, M.; Merino, P.; Nóvoa, X.R. Electrochemical impedance spectroscopy study of the corrosion process on coated galvanized steel in a salt spray fog chamber. *Corrosion* **2000**, *56*, 1220–1232. [CrossRef]
37. Amirudin, A.; Thierry, D. Application of electrochemical impedance spectroscopy to study the degradation of polymer-coated metals. *Prog. Org. Coat.* **1995**, *26*, 1–28. [CrossRef]
38. Lorenz, W.J.; Mansfeld, F. Determination of corrosion rates by electrochemical DC and AC methods. *Corros. Sci.* **1981**, *21*, 647–672. [CrossRef]
39. McIntyre, J.M.; Pham, H.Q. Electrochemical impedance spectroscopy; a tool for organic coatings optimizations. *Prog. Org. Coat.* **1996**, *27*, 201–207. [CrossRef]
40. Cabanelas, I.; Collazo, A.; Izquierdo, M.; Nóvoa, X.R.; Pérez, C. Influence of galvanised surface state on the duplex systems behaviour. *Corros. Sci.* **2007**, *49*, 1816–1832. [CrossRef]
41. Thomas, N.L. The Barrier Properties of Paint Coatings. *Prog. Org. Coat.* **1991**, *19*, 101–121. [CrossRef]

coatings

MDPI

Article

Durability of Acrylic Cataphoretic Coatings Additivated with Colloidal Silver

Massimo Calovi and Stefano Rossi *

Department of Industrial Engineering, University of Trento, 38123 Trento, Italy; massimo.calovi@unitn.it
* Correspondence: stefano.rossi@unitn.it; Tel.: +39-0461-282442

Abstract: In this work, colloidal silver has been added into an acrylic clear cataphoretic bath, evaluating the effect of two different filler amounts on the durability of the composite coatings. The three series of samples were characterized by electron microscopy to assess the possible change in morphology introduced by the silver-based additive. The protective properties of the coatings were evaluated by a salt spray chamber exposure and electrochemical impedance spectroscopy measurements, evidencing the negative effect provided by high amount of silver, which introduced discontinuities in the acrylic matrix. Finally, the durability of composite coatings was studied by exposing them to UV-B radiation, observing a strong phenomenon of silver degradation. Although the coating containing high concentrations of silver demonstrated poor durability, this study revealed that small amounts of silver can be used to provide particular aesthetic features, but also to improve the protective performance of cataphoretic coatings.

Keywords: colloidal silver; cataphoretic deposition process; salt spray chamber; electrochemical impedance spectroscopy; UV-B exposure test

Citation: Calovi, M.; Rossi, S. Durability of Acrylic Cataphoretic Coatings Additivated with Colloidal Silver. *Coatings* **2022**, *12*, 486. https://doi.org/10.3390/coatings12040486

Academic Editor: Mohammadreza Daroonparvar

Received: 14 March 2022
Accepted: 3 April 2022
Published: 4 April 2022

1. Introduction

Among the various deposition processes of organic coatings, cataphoresis represents an innovative method of particular industrial interest. The typical spray process, in fact, can be applied to different substrates, in addition to metal alloy [1–3], such as glass, paper, and cotton fabric [4], but it has always exhibited technical limitations. For example, the control of the thickness of the layer, as well as the geometry of the component to be coated, are critical aspects during spraying. Differently, cataphoresis has shown several interesting features over the years. The electrodeposition process enables the creation of organic layers with high adhesion and corrosion protection properties [5,6], which are also applicable on products with complex geometry [7]. Thanks to the polarization of the sample that occurs during the electrodeposition, the electric field allows one to achieve a uniform coverage on the surface to be coated. Furthermore, from an industrial point of view, cataphoresis is highly appreciated, as it represents a simple, fast, and cost-effective method to produce large-scale industrial coatings, with a low-environmental impact [8].

Thanks to these aspects, cataphoresis is often successfully employed in the automotive industry [9–11], but in recent years it has also aroused particular interest in the research-academic field. In fact, several works have turned to the optimization of the process equipment [12,13] and parameters, such as the curing temperature [14] and the deposition voltage [15], in order to further improve the behavior of the cataphoretic layers. The electrodeposition process has exhibited high versatility, as it has been applied to different metal substrates, such as Zn-Mg alloys [16,17], Ti-Mo alloy [18], aluminum alloys [19–21], and steel alloys [22]. Furthermore, this method has also been evaluated in synergy with different types of pre-treatments [23,24] to improve the adhesion and durability of the coating. Recently, several scientists have focused on the study of the implementation of the cataphoretic organic matrix with different types of fillers, in order to obtain multifunctional composite coatings. The protective performances, for example, have been further improved by adding cerium-based nanoparticles [25,26], or thanks to functionalized graphene oxide flakes [27–30]. Furthermore, the aesthetic features obtainable from cataphoresis by incorporating particular thermochromic pigments were also evaluated [31], which compared the results with the behavior of the typical spray process [32].

A current trend topic in the world of organic coatings is represented by the addition of silver-based fillers [33–35]. This phenomenon is mainly due to the escalation of the SARS-CoV-2 pandemic, which has led scientific research to seek a solution to the spread and transmission of pathogenic microorganisms through surface contamination. The pandemic has in fact highlighted how the continuous surface disinfection is not sustainable, in terms of temporal and economic efforts [36]. The simplest and most effective alternative approach lies precisely in functionalizing the surfaces by adding silver, whose successful biocidal effect has been amply demonstrated over the last 25 years [37–40]. Silver ions, adhering to the cell wall and modifying its permeability, are able to deactivate respiratory enzymes and hinder DNA replication [41,42]. Thus, silver is commonly employed in sensitive health application that requires extreme hygiene, such as artificial implants, surgical instruments, and dental implants [43], or in food and textile industries [40]. Thus, in order to realize multifunctional surfaces with antibacterial features, silver has recently been subjected to much research on its antimicrobial [44], bioactive activities [45], and cytotoxicity features [46].

Silver has been applied to different types of surfaces. Anodized aluminum layers have shown good compatibility with different types of silver-based fillers, confirming excellent antibacterial performance [47–49]. Furthermore, antimicrobial silver has been encapsulated in TiO_2 coatings [50] and TiO_2 nanotubes [51], deposited on cotton fabric [52], and implemented in a Mg-Al layered double hydroxide coating [53].

Consequently, the world of paints has also been recently affected by these studies [54–58]. For example, Lateef et al. [59] investigated the efficiency of silver nanoparticles (AgNPs) as an antibacterial additive for enamel paint, while Deyà et al. [60] and Barberia-Roque et al. [61] studied the bactericidal effect of silver in acrylic water-based paint. Similarly, Liu et al. [62] explored the in situ synthesis of AgNPs in transparent PVA films for antibacterial purpose. The biocidal activity of AgNPs in polyurethane coating was assessed by Bechtold et al. [63], who confirmed the good efficiency of silver against bacteria but, at the same time, the non-satisfactory resistance against fungi. On the other side, Asafa et al. [64] reported the influence of AgNPs on physical and mechanical properties of emulsion paint, comparing the results with the behavior of ZnO and Fe_2O_3 additives. Finally, the antibacterial properties of silver in emulsion paint was also assessed by Ohashi et al. [65], who evidenced a better paintability and transparency of silver–cytokinin complex filler for emulsion paint compared to a commercialized paint filler.

However, despite the great interest of the paint industry in silver-based additives, the literature does not offer any study on the synergy between cataphoresis and silver. The high process yields of cataphoresis, combined with the excellent covering power of the electrodeposited layers and their compactness, makes this method one of the most promising for the production of organic matrix composite coatings incorporated with silver-based filler. Thus, the purpose of this work is to evaluate the effect of colloidal silver on the durability of acrylic cataphoretic coatings. Two different amounts of colloidal silver were added into the cataphoretic bath, to assess how the silver concentration influences the coatings' performance. The study employed an acrylic top-coat cataphoretic bath, as it did not contain pigments and additives that could influence the effect and behavior of colloidal silver.

The samples were characterized by means of the scanning electron microscope (SEM), to observe the morphology of the deposited composite coatings by studying the influence of the silver additive. The effect of the amount of silver on both corrosion resistance performance and behavior in the aggressive environment of the acrylic-based coatings were assessed by subjecting the samples to electrochemical impedance spectroscopy measurements (EIS) and exposing them in the salt spray chamber, respectively. Finally, the durability of the samples and consequent aesthetic performance were evaluated with the UV-B exposure.

2. Materials and Methods

2.1. Materials

The colloidal silver (65–75 wt.% Ag) and acetone were supplied from Sigma-Aldrich (St. Louis, MO, USA) and used as received. The 40 × 70 × 2 mm^3 carbon steel substrate (Q-panel type R (0.15 wt.% C-Fe bal.) was supplied by Q-lab (Westlake, OH, USA). The cataphoretic bath Arsonkote 202 Acy Cata W202 × 30 was provided by Arsonsisi (Milan, Italy). This industrial product is based on modified acrylic resin, with glycols employed as the solvent. It possesses a dry residue (after 2 h at 120 °C) equal to 28–30%, and a specific weight of 1.00–1.20 g/cm^3.

2.2. Deposition of Acrylic–Ag Coatings

The carbon steel substrates were subjected to a degreasing process in acetone, supported by ultrasonic agitation, followed by a sandblasting step with corundum powder (0.2 mm diameter, 70 mesh). The sandblasting process introduced a roughness [Ra] equal to 3.05 ± 0.22 µm in the steel substrates. Finally, the steel plates were further degreased in acetone, in order to remove possible traces of contamination.

The deposition setup was set in a 200 mL acrylic cataphoretic bath. A 300 × 50 mm^2 flat stainless steel plate, used as the anode, was placed in front of the sample, at a distance of 80 mm. As colloidal silver granules are soluble in water, they were added directly to the cataphoretic bath, because it is water-based. Two different amounts of colloidal silver were added into the bath, which was stirred for 30 min with an ultrasound probe, to facilitate the complete dissolution of the colloidal silver granules. Both the two sets of samples were deposited, keeping constant all the process parameters, to better compare the coating behavior. The electrodeposition was carried out at a voltage of 75 V for 120 s, followed by a curing step in an oven at 140 °C for 45 min, as suggested by previous studies [29,30]. The sample production process is outlined in Figure 1.

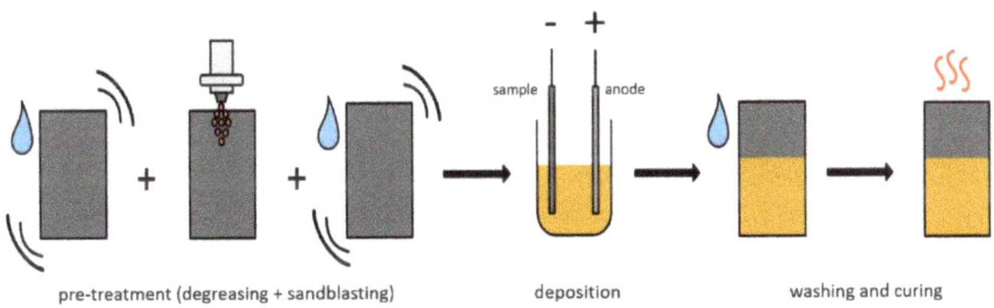

Figure 1. Scheme of the sample production process.

The performances of the two series of acrylic–Ag samples were compared with the behavior of a pure acrylic layer, free of filler. The three samples series are summarized in Table 1 with the sample nomenclature.

Table 1. Labeling of samples with different colloidal silver concentration.

Bath	Colloidal Silver Concentration (wt.%)	Sample Nomenclature
Clear coat	0.00	A
	0.05	A1
	0.10	A2

The two silver concentrations, shown in Table 1, were chosen because they achieved enough silver quantity in the coating in order to guarantee good antibacterial performances, as highlighted in the literature [48]. The real amount of silver detected inside the two coatings A1 and A2 are introduced later, in paragraph 3.1.

2.3. Characterization

The optical stereomicroscope Nikon SMZ25 (Nikon Instruments, Amstelveen, the Netherlands) and low vacuum scanning electron microscope SEM JEOL IT 300 (JEOL, Akishima, Tokyo, Japan) were employed to analyze the surface morphology of the coatings, with the aim of verifying whether the introduction of colloidal silver has influenced the electrodeposition process, modifying its yield and introducing defects in the acrylic matrix. An energy-dispersive X-ray spectroscopy (EDXS, Bruker, Billerica, MA, USA) analysis has also been carried out in order to map the silver distribution in the coatings. Finally, the thickness of the three series of composite coatings was measured by means of the Phynix Surfix digital thickness gauge (Phynix, Neuss, Germany).

The samples were exposed in a salt spray chamber (Ascott Analytical Equipment Limited, Tamworth, UK) for 500 h, following a ASTM B117-11 standard [66] (5 wt.% sodium chloride solution), to assess the corrosion protection behavior of the coatings in a particular aggressive environment. The adhesion of the coatings was evaluated by means of a mechanical cut made on the surface of the samples, analyzing the possible detachment and water uptake phenomena.

The protective features of the composite coatings were studied by means of Electrochemical Impedance Spectroscopy (EIS) measurements, carried out with a potentiostat Parstat 2273 (Princeton Applied Research by AMETEK, Oak Ridge, TN, USA) with the software PowerSuit ZSimpWin (version 2.40) and applying a signal of about 15 mV (peak-to-peak) amplitude in the 10^5–10^{-2} Hz frequency range. The cell setup mas composed of an Ag/AgCl reference electrode (+207 mV SHE) and a platinum counter electrode, immersed in the 3.5 wt.% sodium chloride solution. The samples were kept immersed in the test solution for a total of 500 h, with a testing area equal to 6.15 cm^2.

Finally, UV-B exposure tests were carried out at 313 nm for 500 h, using an UV173 Box Co.Fo.Me.Gra (Co.Fo.Me.Gra, Milan, Italy) and following the ASTM G154-16 standard [67]. The chemical–physical degradation of the coatings was analyzed with a Konica Minolta CM2600d, spectrophotometer (Konica Minolta, Chiyoda, Tokyo, Japan) using a D65/10° illuminant/observer configuration in SCI mode, evaluating visual-aesthetical variations.

3. Results

3.1. Coating Morphology

The deposition of the acrylic–Ag composite coatings took place by adding the clear cataphoretic bath with colloidal silver. The silver-based additive is in the form of a greyish powder, with a very variable size, as shown in Figure 2. In fact, the size of the granules varies from a few µm, up to over 200 µm.

However, this aspect seems not to be relevant for the electrodeposition process, as colloidal silver powders are completely soluble in water. Consequently, once added to the cataphoretic bath, composed of 60 wt.% of water, the powders did not reveal macroscopic aggregation phenomena. Previous works have highlighted the strong criticalities in adding fillers in cataphoretic baths: the reaction with the polymeric matrix often results in aggregation phenomena of the fillers with consequent defects in the coatings [29,30]. Otherwise, colloidal silver apparently represents an ideal candidate as an additive in cataphoretic depositions.

Dissolving in water, the colloidal silver powders create a yellow-orange color solution. Consequently, the color of the cataphoretic bath also varies according to the quantity of the introduced colloidal silver. The hue of the electrodeposition bath, from transparent, varies towards an orange tone, becoming darker and darker with the increase in the colloidal silver amount. Figure 3 faithfully represents this phenomenon: the three types of samples studied in this work exhibit different aesthetic features due to the presence of colloidal silver. Sample A has a clear coating of pure acrylic matrix, free of additives. On the other hand, samples A1 and A2 reveal a yellow and orange coating, respectively, the tones of which strongly depend on the concentration of the silver-based filler. The coloring of the composite layers appears homogeneous and constant over the entire surface, proof of the high aesthetic efficiency and covering power of cataphoresis [32].

Figure 2. Colloidal silver acquired by SEM.

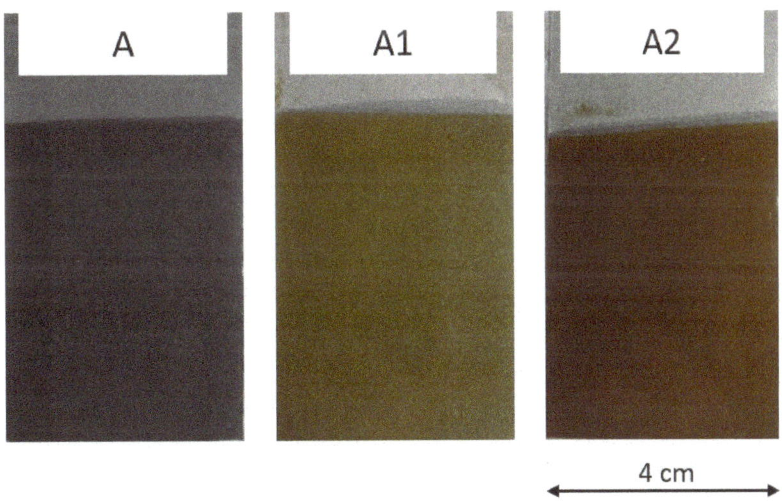

Figure 3. Photograph of the three as-made sample series, where A represents the coating of pure acrylic matrix, while A1 and A2 were deposited by introducing 0.05 wt.% and 0.10 wt.% of colloidal silver in the cataphoretic bath, respectively.

Although the three series of samples exhibit different appearance from each other, the thickness of the coatings is comparable. Table 2 shows the coatings thickness measured with the digital thickness gauge: the values constitute the average of 50 measurements performed on 5 samples (10 measurements per sample) for each series.

Table 2. Coatings thickness of the three series of samples.

Sample	Thickness [μm]	St. Dev. [μm]
A	22.2	0.8
A1	23.4	1.0
A2	21.7	1.3

The obtained thickness values are in accordance with the usual outcome achieved with cataphoretic process, as confirmed by the literature [9]. The results suggest that the addition of colloidal silver to the cataphoretic bath did not affect the process yield of the electrodeposition. The samples show a compact and homogeneous coating, free from macroscopic defects. This outcome reinforces the previous assumptions relating to the efficient use of colloidal silver as an additive in cataphoretic baths. Furthermore, the introduction of colloidal silver in the acrylic bath did not affect the leveling power of the cataphoretic process, as all the three series of coatings exhibit a particularly smooth coating, whose roughness [Ra] is comparable and equal to 0.25 ± 0.03 μm.

Thus, the three series of coatings were observed by SEM, in order to better analyze their morphology and highlight possible microscopic defects. Figure 4 shows the micrographs of the surfaces of the three samples, acquired by SEM.

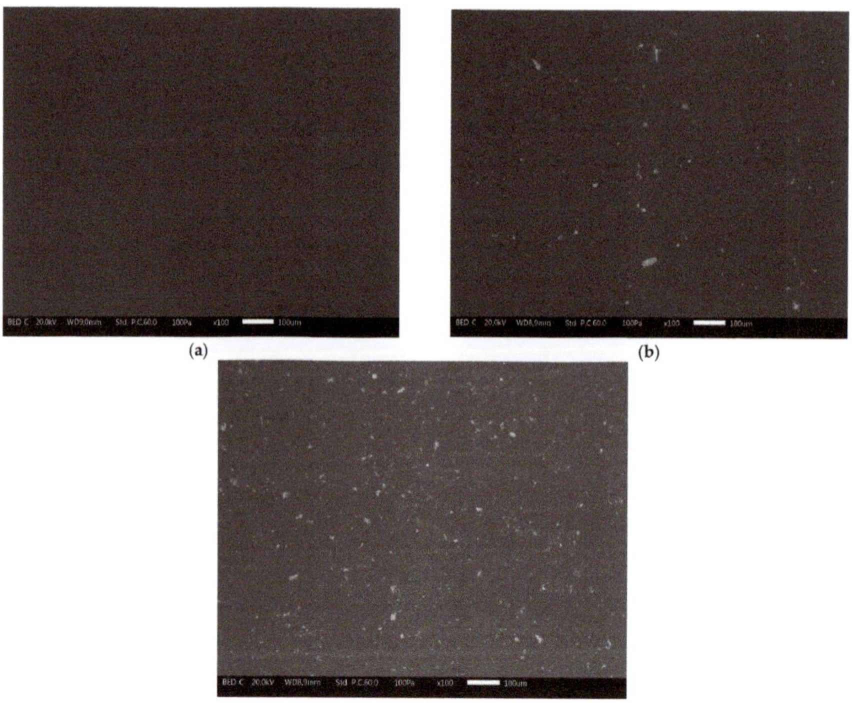

(a) (b) (c)

Figure 4. SEM micrograph of the surface of sample A (**a**), sample A1 (**b**) and sample A2 (**c**), respectively.

Coating A (Figure 4a) appears completely homogeneous. Morphological defects, such as bubbles in the acrylic matrix, cannot be appreciated. In contrast, the coatings of samples A1 and A2 (Figure 4b,c, respectively) reveal different light-colored traces homogeneously distributed on their surface. The amount of these traces increases from sample A1 to sample A2, suggesting that they are closely related to the addition of colloidal silver to the acrylic bath. Despite the presence of this phenomenon, no appreciable defects are observed even in coatings A1 and A2. Therefore, although silver is conductive, once introduced into the cataphoretic bath it does not influence the electrodeposition process through the development of eddy currents, as observed, for example, with graphene-based filler [29].

In order to confirm the nature of the clear traces, the samples were examined by EDXS investigation. The EDXS analysis revealed a concentration equal to 1.4 wt.% and about 3 wt.% of silver on the surface of the samples A1 and A2, respectively. Figure 5 shows, as an example, a micrograph of the surface of the sample A1, with the respective EDXS map of the Ag element. Indeed, the clear traces observed by SEM are representative of the presence of silver. Although silver is effectively distributed over the entire surface of the coatings in a homogeneous way, in some cases it is possible to notice a thickening of silver whose dimensions exceed 20 μm. The morphology of these clusters differs from the powders shown in Figure 2. Consequently, it is likely that the colloidal silver actually dissolved in water, but was subsequently subjected to thickening phenomena in the acrylic matrix. In fact, colloidal silver itself is not compatible with polymeric matrices, if not first effectively dissolved in water. It is therefore probable that the interaction between the silver–water solution and the acrylic resin in the cataphoretic bath has caused the undesired phenomena of silver re-accumulation. This, in fact, represents the only real noteworthy defect due to the addition of colloidal silver.

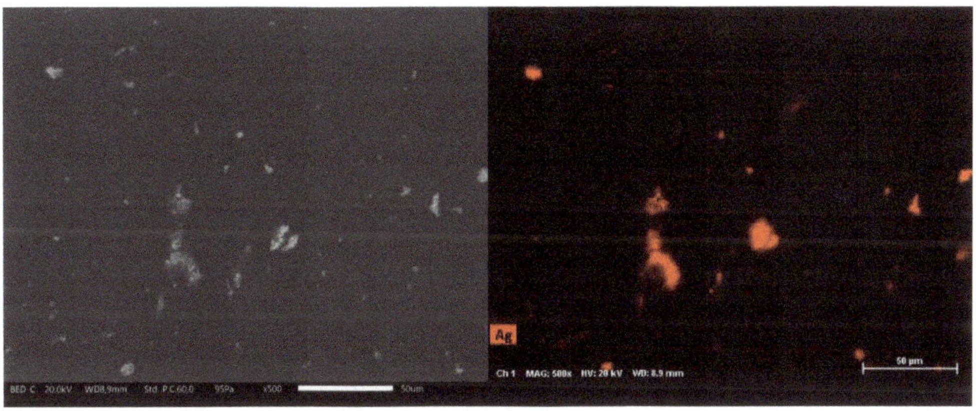

Figure 5. SEM micrograph of the surface of sample A1, with relative EDXS map of Ag element.

Thus, the effect of this phenomenon on the protective performances of the acrylic matrix was subsequently evaluated by exposing the coatings in aggressive environments and by means of suitable electrochemical techniques.

3.2. Salt Spray Chamber

The three series of samples were exposed in the salt spray chamber for a total of 500 h. During the first 100 h of exposure the coatings were observed every 24 h, after which they were monitored every 100 h, until the end of the analysis. An artificial cut of 1 mm width was realized on the surface of the coatings, to create a mechanical defect in the cataphoretic layer, forcing the development of corrosive phenomena at the substrate-coating interface. Thus, the evolution of blisters, due to water uptake from to the artificial notch, and the coatings' adhesion level were visually evaluated [68].

Figure 6 shows the evolution of the coatings' degradation during the exposure in the salt spray chamber. All three series of coatings exhibit a halo around the central notch after 100 h of exposure. This phenomenon, together with the development of blisters, is symptomatic of water absorption in correspondence of the artificially introduced defect. The samples exhibit a comparable behavior, except for the greater blister evolution observed in coating A2. At the end of the test (500 h), the entire surfaces of the three coatings are covered with blisters. However, it must be considered that cataphoresis is normally used as a primer or as a finishing layer in non-aggressive environments, while the exposure test in salt spray chamber exhibits a high degree of aggressiveness. The absorption of the aggressive solution within the cataphoretic layers is associated to their adhesion levels: the greater the distance of the blister from the central notch, the lower the adhesion of the coatings. Again, it is difficult to differentiate between the behavior of the three samples, as the penetration of the test solution caused the development of blisters over the entire surface of the coatings.

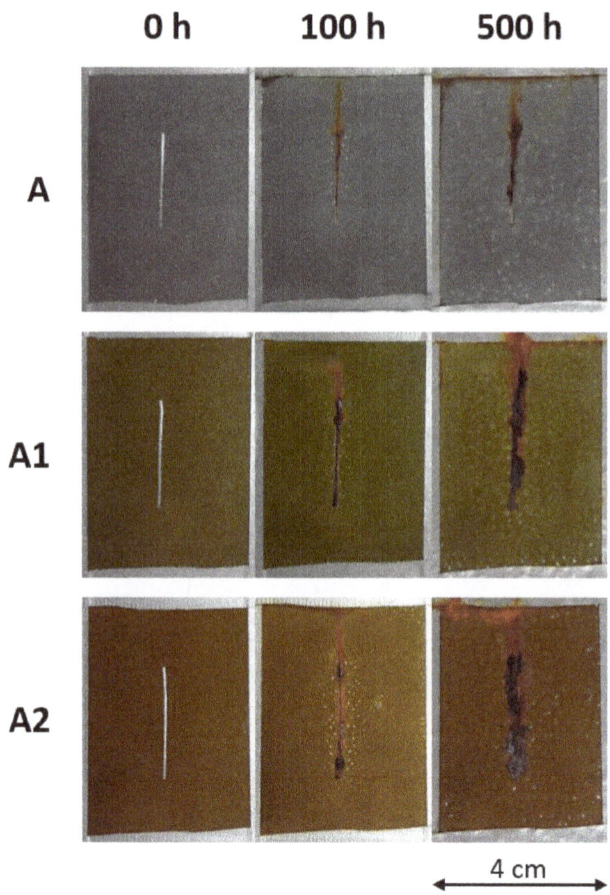

Figure 6. Coating degradation during sample exposure in salt spray chamber.

To better observe the phenomena that took place close to the artificial notch, the corrosion products developed on the coatings were removed by stirring the samples in citric acid solution (pH = 3) for 2 h. Thus, the samples were analyzed by stereomicroscope observations, as shown in Figure 7. According to the standard [66], an organic coating with good adhesion to the substrate should show maximum detachment from the notch area of about 1000 μm. Following this definition, all three series of coatings possess excellent adhesion, typical of a cataphoretic layer, as they did not exhibit clear detachment of the coating caused by water absorption or development of corrosion products. However, in some spots it is possible to appreciate the detachment of the layers from the metal substrate in proximity of the notch: this phenomenon is represented by lighter blisters (clearly evident above all in Figure 7a,b). This occurrence sometimes exceeds the limit of 1000 μm, but it is a detachment of the layer strictly linked to the evolution of the blister, therefore not fully considered by the standard [66].

(a)

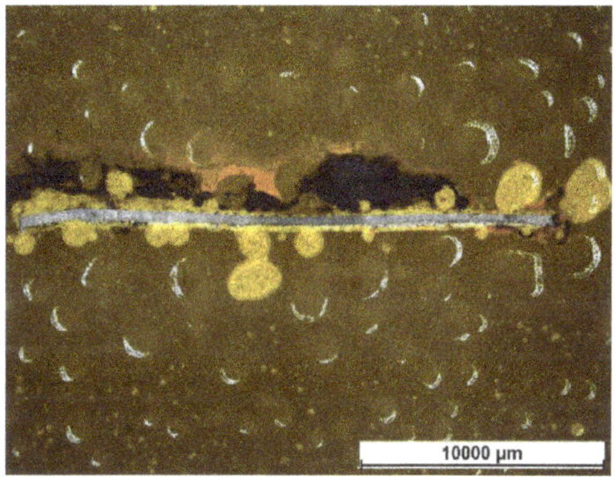

(b)

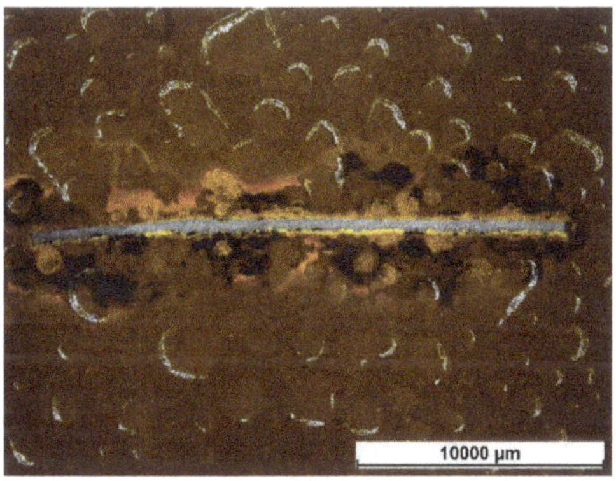

(c)

Figure 7. Samples A (**a**), A1 (**b**), and A2 (**c**) appearance after 500 h of exposure in the salt spray chamber.

The density of blisters, rather than their average size, represents the aspect that most differentiates the three series of samples. As can be observed from the three images in Figure 7, the presence of colloidal silver seems to cause an increase in the density of blisters that develop following the exposure of the coatings to the aggressive environment of the salt spray chamber. Furthermore, this phenomenon is even more marked after increasing the amount of silver present in the coatings. Referring to the standard [68], the three series of samples have blisters of size 1–3. However, the density of the blisters increases from grade 3 of sample A, to grades 4 and 5 of samples A1 and A2, respectively. Consequently, silver seems to favor the adsorption of water into the cataphoretic coating, as it represents a point of discontinuity in the acrylic matrix. Figure 8, acquired with the optical stereomicroscope, is exemplar of the high blister density exhibited by sample A2 at the end of the test. The polarizing filter of the microscope made it possible to highlight the morphology of the blisters, which in some cases exceed 2000 μm in size.

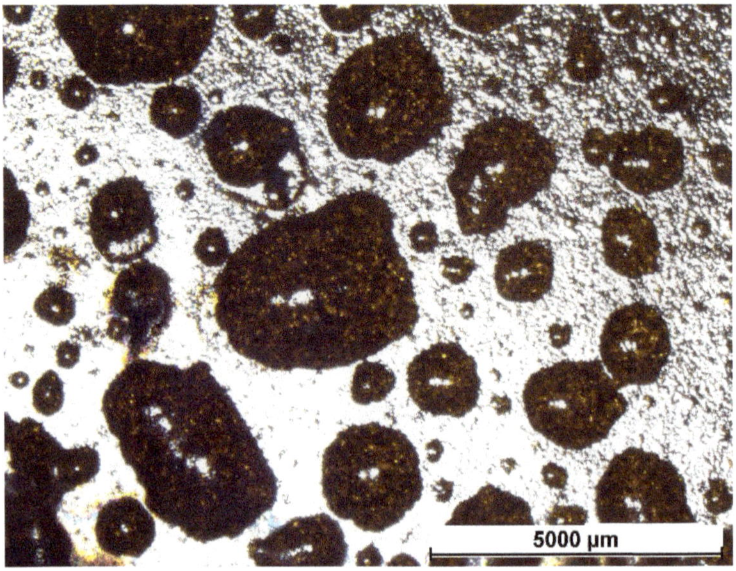

Figure 8. Detail of the blisters in sample A2 observed with the optical stereomicroscope at the end of the test.

Therefore, although silver does not introduce apparent defects in the cataphoretic layer, it influences the behavior of the coating for long exposures in aggressive environments. The difference between the results expressed by sample A and sample A1 is almost negligible, while the greater quantity of silver present in the coating A2 seems to play a key role in the durability of the acrylic matrix. To better study the protective characteristics of the three types of coatings evaluating the contribution of silver, the samples were subjected to Electrochemical Impedance Spectroscopy (EIS) measurements.

3.3. Electrochemical Impedance Spectroscopy Measurements

The EIS measurements are widely employed for the estimation of the protective feature of organic coatings, as this technique is able to provide useful details about the corrosion resistance properties of the layer, such as its defectiveness or degree of adhesion [69].

For example, the evolution of the Bode impedance module measured at low frequencies (10^{-2} Hz), defined as $|Z|_{(0.01)}$, can be monitored to get a preliminary quantitative estimation of the degree of protection provided by the coating. The literature defines a minimum value of $|Z|_{(0.01)}$, equal to 10^6 $\Omega\cdot cm^2$, below which the coating does not ensure certain protective performances [70,71]. Thus, the protective behavior of the three series of coatings was characterized monitoring the evolution over time of their impedance module $|Z|_{(0.01)}$ over time. Figure 9 shows the variation of the parameter $|Z|_{(0.01)}$ during the exposure of 500 h of the samples to the EIS test solution.

The three coatings exhibit different behaviors during the experiment. Sample A reveals a relatively high initial value of $|Z|_{(0.01)}$, which drops, however, by an order of magnitude during the first 4 h of exposure to the test solution. Subsequently, a continuous decrease is observed, with slower speed, until reaching a plateau towards the 340 h of testing. The value of this plateau is very close to the limit of 10^6 $\Omega\cdot cm^2$, suggesting that the acrylic matrix no longer provides the right protective guarantees.

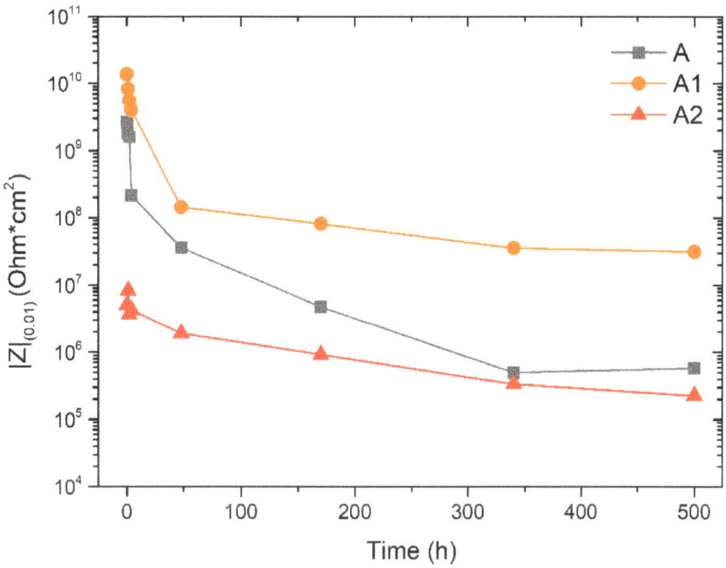

Figure 9. Bode impedance modulus $|Z|_{(0.01)}$ evolution with time.

Similarly, sample A1 also shows a sudden decrease in the value of $|Z|_{(0.01)}$ during the first hours of testing. However, the curve quickly settles on high values, equal to about 10^8 $\Omega \cdot cm^2$, keeping this plateau almost unchanged until the end of the test. The improvement of the protective performance of organic coatings by adding silver has already been observed in the literature, under certain conditions [72,73]. This phenomenon can be associated with a decrease in the intrinsic porosity of the acrylic matrix in the presence of limited quantities of silver. By acting as a very low permeability filler, silver is able to reduce the overall porosity of the composite coating. Consequently, the reduced permeability of the coating translates into greater protective performances, expressed by higher values of the impedance module $|Z|_{(0.01)}$.

However, as already observed during exposure in salt spray chamber, too high quantities of silver introduce structural changes in the matrix, increasing its permeability [74]. This phenomenon affects the result of the impedance measurements of sample A2: the trend of $|Z|_{(0.01)}$ immediately starts from lower values than the other two samples, A and A1. In this case, by increasing the permeability of the acrylic matrix, silver reduces its protective properties right away. Throughout the test, sample A2 always shows lower performance than coating A, of the pure acrylic matrix. Having reached 500 h of exposure, $|Z|_{(0.01)}$ comes dangerously close to values of 10^5 $\Omega \cdot cm^2$, evidencing poor protective performance. These results are in agreement with the salt spray test, confirming the criticalities found in the composite coating A2 containing high amount of silver.

The three series of samples showed a strong decrease in $|Z|_{(0.01)}$ mainly during the first 48 h of exposure to the test solution. Indeed, the first hours of immersion are the most significant, as typically the organic coatings undergo solution absorption phenomena. To better analyze the behavior of the three samples and explain the results exhibited in Figure 9, the evolution of the Nyquist diagrams during the first 48 h of the test is displayed in Figure 10. The three images present a box with a focus on low values of Z' and $-Z''$, to highlight the details of the diagrams acquired after 4 h and 48 h of immersion in the test solution.

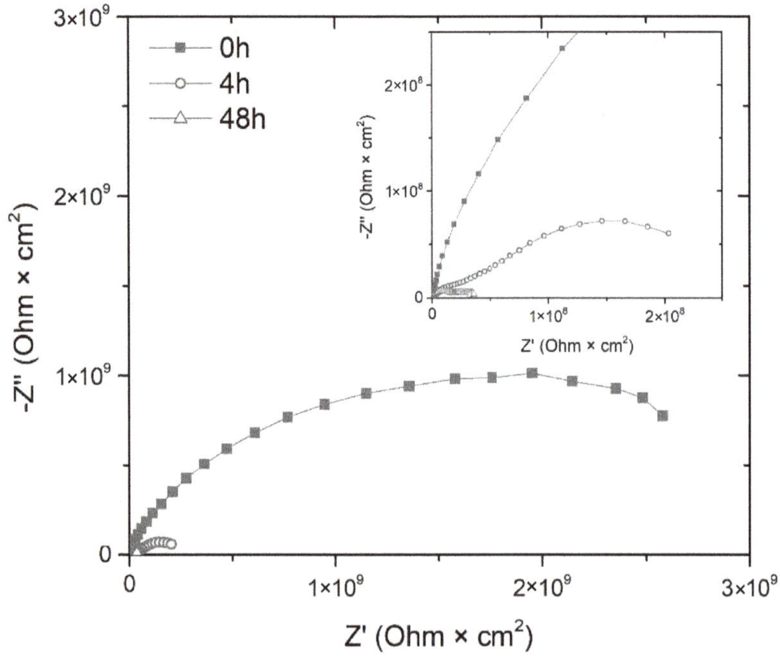

(a)

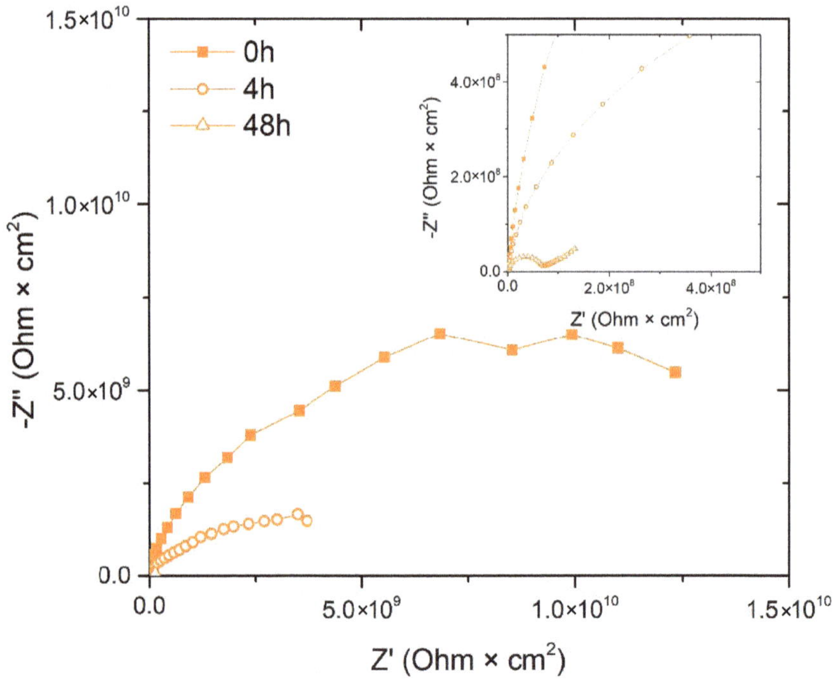

(b)

Figure 10. *Cont.*

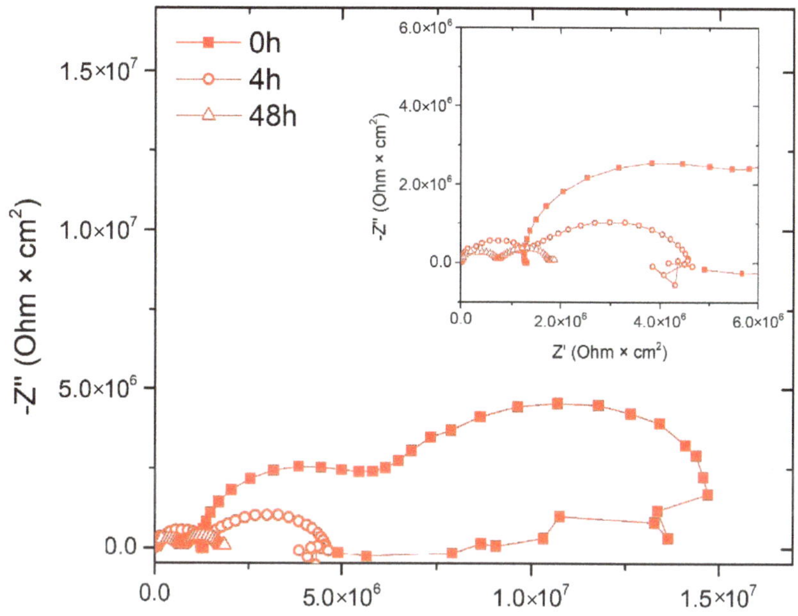

(c)

Figure 10. Nyquist diagram evolution during the first 48 h of test of the samples A (**a**), A1 (**b**), and A2 (**c**), respectively.

The Nyquist diagrams of sample A (Figure 10a) reveal a rapid decrease in the protective performance of the coating, as both the resistive and capacitive contributions of the acrylic layer are reduced over time. The diagrams represent two-time constants, which are more and more evident as the exposure to the test solution proceeds. The presence of two-time constants indicates the occurrence of corrosion phenomena: the time constant at low frequencies represents, in fact, the dissipative phenomena that occur at the coating-substrate interface, as a consequence of the rapid absorption of the solution. Thus, the acrylic matrix is confirmed to possess intrinsic poor protective features.

The diagrams in Figure 10b (sample A1) exhibit a similar trend, even if the two-time constants can only be appreciated after 48 h of measurements. The dissipation phenomena below the coating seem to be postponed over time, suggesting the barrier effect introduced by silver, which acts as a filler for the porosity of the acrylic matrix. This phenomenon translates into a better protective performance of sample A1 throughout the test.

Finally, the Nyquist diagrams of sample A2 (Figure 10c) provide the most significant information. The two-time constants are already evident in the diagram acquired at the start of the test, which shows two distinct semicircles. However, the second semicircle, representative of the dissipative phenomena due to the composite coating, tends to close in on itself, with the curve returning to lower Z′ values. This is a typical behavior of systems in continuous evolution during the measurement itself: it represents the tendency of the composite coating to absorb a considerable amount of solution, so as to make the data acquisition pretty unstable. Consequently, the impedance values, already low at the start of the measurements, tend to decrease quickly until they reach and exceed the safety threshold of 10^6 $\Omega{\cdot}cm^2$, as shown in Figure 9.

The behavior of the three series of samples is well represented by the images in Figure 11, representative of the surfaces subjected to the EIS measurements after 500 h of immersion in the test solution. The surfaces of the three samples exhibit some small dark corrosion products, but still differ in the density of blisters developed during the experiment. More than half of the surface of the coating A reveals the presence of blisters of different sizes. The addition of 0.05 wt.% of colloidal silver results in an improvement in the durability of the acrylic matrix, as the development of blisters is limited. Finally, the poor results of the EIS measurements relating to sample A2 can be appreciated from the appearance of the coating, whose surface at the end of the test is completely covered with blisters.

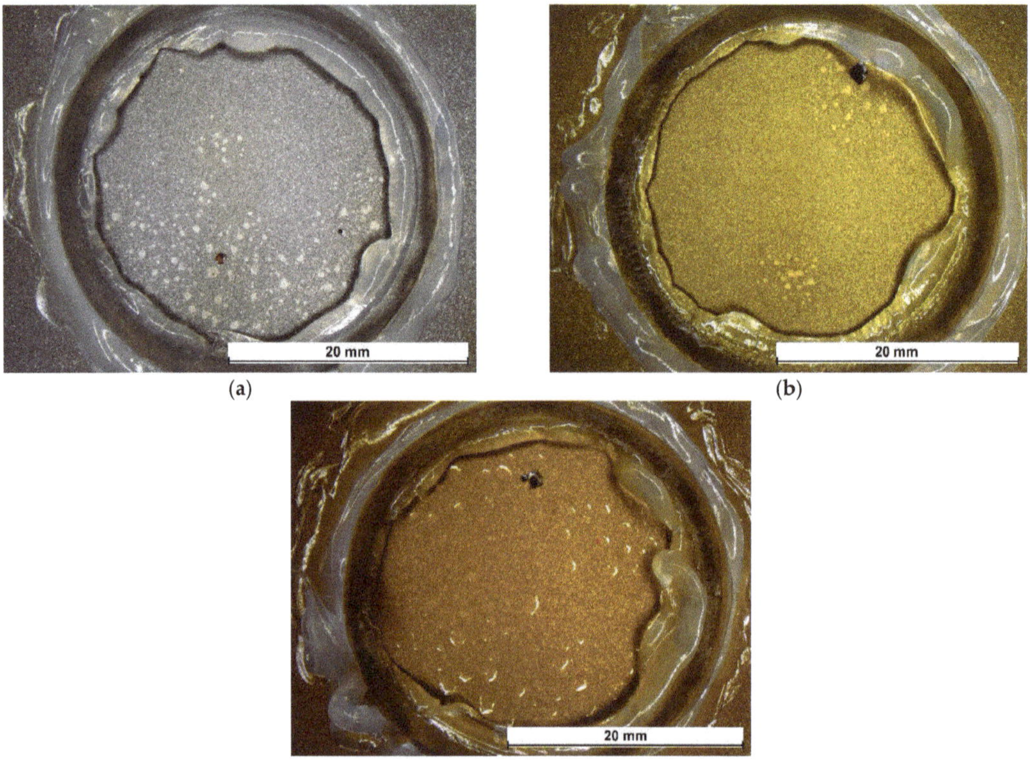

(a) (b) (c)

Figure 11. Surface of sample A (**a**), A1 (**b**) and A2 (**c**), after 500 h of immersion in test solution.

Therefore, the simple observation of the surfaces of the three coatings provides realistic information on their protective performance, in agreement with the results of specific techniques such as the electrochemical impedance spectroscopy measurements.

As both the test of exposure in salt spray chamber and the EIS measurements exhibit comparable results, the positive contribution introduced by small quantities of colloidal silver has been confirmed. However, the addition of Ag can take on negative connotations if a limit concentration is exceeded, reducing the protective performance of the acrylic matrix.

3.4. Exposure to UV-B Radiation

Ultraviolet radiation can affect the aesthetic features of organic coatings, causing their chemical degradation. Thus, UV-B exposure is a procedure widely employed for the assessment of the durability of organic coatings, which must ensure a long outdoor service life [75,76].

While the limited effect of solar radiation on acrylic resins is well known [77,78], recent studies highlighted the degradation phenomena that silver can undergo when exposed to UV light, in terms of retention, dissolution, and oxidative aging [79,80]. The extent of these phenomena can also be exacerbated due to the combination with high temperatures [81]. Therefore, the samples were exposed for 500 h to UV-B radiation at 50 °C, evaluating the change in aesthetic features of the cataphoretic coatings.

Figure 12 shows the total color variation, ΔE, every 100 h of UV-B light exposure. According to the ASTM E308 (2018) standard [82], ΔE is calculated as follows:

$$\Delta E = [(\Delta L^*)^2 + (\Delta a^*)^2 + (\Delta b^*)^2]^{1/2} \tag{1}$$

where the colorimetric coordinates L*, a* and b* represents the lightness (0 for black and 100 for white objects), the red–green coordinate (positive values are red, negative values are green), and the yellow–blue coordinate (yellow for positive values, blue for negative values and 0 as neutral parameter), respectively.

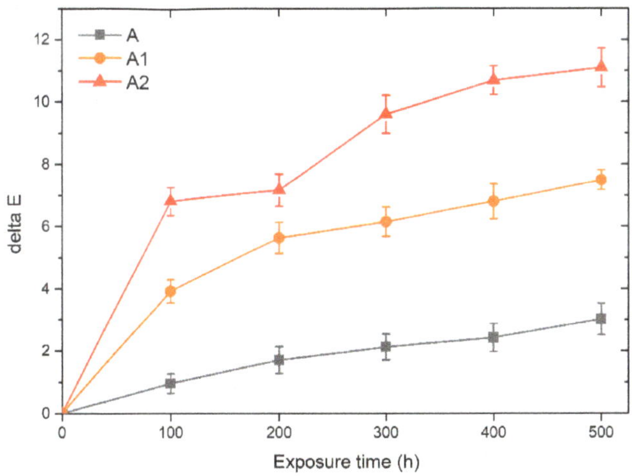

Figure 12. Total color variation during UV-B exposure.

ΔE = 1 is considered the threshold value for which the color difference is imperceptible to the human eye [83]. Consequently, the coating A exhibited a limited change in appearance. The color of the acrylic matrix varied constantly during the test; however, it always remained very close to the initial coordinates. Indeed, after 500 h of exposure to the UV-B light, coating A revealed a value of ΔE approximately equal to three, confirming the good durability of the acrylic matrix against solar radiation.

Otherwise, the two silver-containing coatings underwent a noticeable color change. Silver proves to suffer particularly from exposure to UV-B rays, as sample A1 and sample A2 exhibited a sudden and not negligible change in color. After only 100 h of exposure, the color change reaches values over half of the total ΔE observed during the entire test. The greater the quantity of silver contained in the acrylic matrix, the greater the variation in the aesthetic features of the coating, but also the more intense the degradation due to the exposure to UV-B radiation.

Figure 13 compares the evolution of the appearance of the three coatings during the UV-B exposure test. While the degradation of coating A cannot be appreciated, the human eye easily recognizes the tendency of the coatings containing silver to become darker.

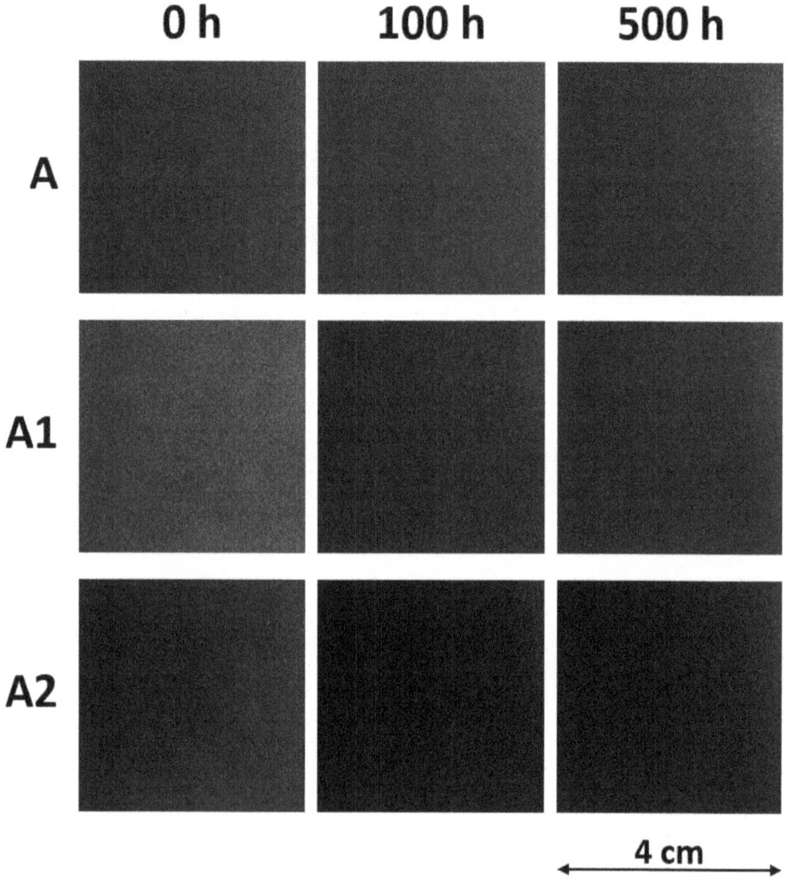

Figure 13. Evolution of the appearance of the coatings during the exposure to UV-B radiation.

However, this phenomenon is due to the variation of the values of the coordinates a* and b*, as shown in the Table 3, rather than linked to a decrease in the brightness (L*) of the coating.

Table 3. Evolution of the colorimetric coordinates of the samples during the exposure to UV-B radiation.

		A			A1			A2	
Time [h]	**L***	**a***	**b***	**L***	**a***	**b***	**L***	**a***	**b***
0	48.26	0.11	2.26	43.61	0.77	17.96	38.01	6.3	19.32
100	47.43	−0.18	2.64	42.48	2.95	14.88	34.97	10.23	14.65
200	47.18	−0.44	3.47	41.36	2.76	13.21	34.65	9.2	13.71
300	47.02	−0.47	3.89	41.06	2.68	12.71	33.41	9.83	11.68
400	47.37	−0.47	4.44	41.06	2.31	11.86	34.4	7.59	9.36
500	46.79	−0.4	4.85	40.95	2.2	11.12	35.89	6.13	8.45

Before exposure to UV-B radiation, the addition of silver in the acrylic matrix caused a decrease in the brightness of the coating (lower L* values) and an increase in the coordinates a* and b*, representative of shades tending to red and yellow, respectively. Subsequently, the parameter b*, even more than the coordinate a*, varies towards lower values, decreasing the yellow shade of the composite coating. During the whole test, sample A1 and sample A2 showed total Δb* of approximately 7 points and 11 points, respectively. Definitively, this variation in the values of the color coordinates, with consequent darkening of the appearance of the coating, is mainly due to the oxidation phenomena [84] that silver undergoes during UV-B exposure.

The aesthetic variation of the silver-containing coatings is also closely linked to a morphological degradation of the composite layers. Figure 14 reveals the surface of sample A1 and sample A2 observed with the optical microscope at the end of the experiment, after 500 h of exposure to UV-B radiation. Both coatings show the development of some blisters within the acrylic matrix. This phenomenon is intensified in sample A2, suggesting that it has been caused by the presence of silver. Dissolution of silver, due to UV exposure, is supposed to have created these voids and blisters in the acrylic matrix. In addition to a purely aesthetic aspect, this phenomenon represents a critical defect in the coating, as the permeability of the acrylic matrix is further exasperated, compromising its durability. It must be considered that the exposure to UV-B radiation represents a particularly aggressive test, not fully simulating the exposure to sunlight, in terms of frequency and intensity of the radiation. However, the degradation of silver inevitably also leads to a decrease in the protective performance of the cataphoretic coating, suggesting the application of this type of coating only on products designed for indoor applications.

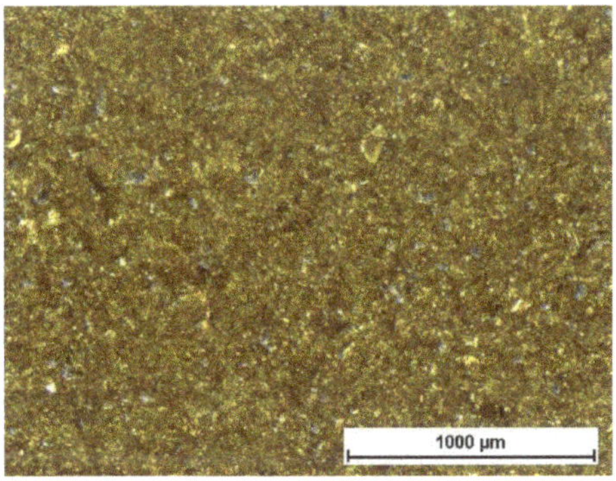

(a)

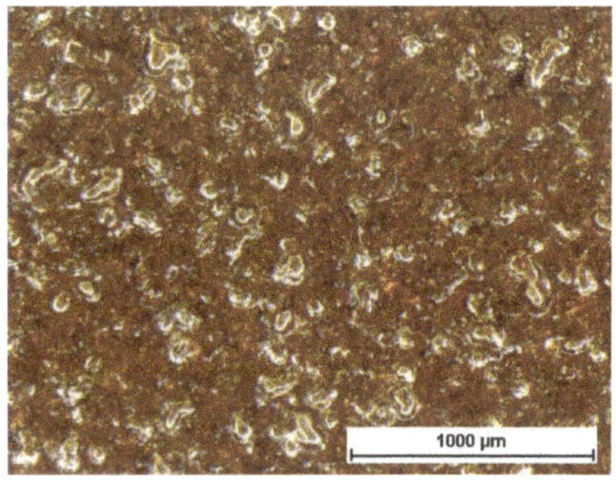

(b)

Figure 14. Detail of the blisters in samples A1 (**a**) and A2 (**b**), observed with the optical stereomicroscope after 500 h of exposure to UV-B radiation.

The silver concentration detected by EDXS analyses in sample A1 and sample A2, equal to 1.4 wt.% and about 3 wt.%, respectively, provides high guarantees on the antibacterial performance of both types of coatings. In fact, other works have confirmed the high antibacterial functionality of silver in low concentrations, even lower than those evidenced by the two composite coatings [42,48]. While coating A2 possesses high defects, with consequent poor durability performance, the quantity of silver present in coating A1 allows both to offer guarantees from the point of view of antibacterial properties, and to increase the protective performance of the acrylic matrix. Consequently, coating A1 can be used in various application fields where good durability is required, in combination with effective antibacterial features.

4. Conclusions

The effect of the addition of colloidal silver on the durability of the cataphoretic coating has been evaluated in this work. The introduction of colloidal silver in the cataphoretic bath caused a significant change in the aesthetic features of the coating, whose color tended towards yellow and orange, the tones of which strongly depended on the concentration of the silver-based filler. Colloidal silver, completely soluble in water, did not introduce defects in the coating and did not influence the electrodeposition process. However, the reaction with the acrylic resin caused limited agglomerations of the silver ions to form discontinuities in the acrylic matrix.

These discontinuities can favor an increase in the permeability of the acrylic matrix, reducing its protective performance. The exposure of the samples in the salt spray chamber highlighted the high absorption of the test solution by the coating containing a high amount of silver. The presence of silver did not directly influence the adhesion of the coating, but favored the water uptake in the acrylic matrix, with consequent development of the high blister density and corrosion products at the interface with the metal substrate.

This phenomenon was confirmed by the electrochemical impedance spectroscopy measurements, which demonstrated a sudden absorption of the test solution by the coating A2, which, in turn, exhibited a decrease in the protective performance of the acrylic matrix. However, the limited amount of silver present in sample A1 resulted in an improvement in the protective behavior of the cataphoretic coating, as the colloidal silver acted as a filler for the porosity of the acrylic matrix. As a matter of fact, after 500 h of exposure to the test solution, sample A1 exhibited a $|Z|(0.01)$ value equal to 108 $\Omega\cdot cm^2$, two orders of magnitude higher than the pure acrylic matrix of coating A.

Finally, the exposure of the samples to UV-B radiation confirmed the high reactivity of silver-based fillers when subjected to sunlight. The rapid oxidation of silver caused a darkening of the appearance of the coating, as evidenced by the colorimetric measurements, but is also visually appreciable. At the same time, the dissolution of silver led to the development of blisters in the acrylic matrix, especially in sample A2, which was produced with a high amount of silver.

Ultimately, this work highlights the merits and criticalities in the use of colloidal silver as a filler in acrylic cataphoretic baths. High concentrations of colloidal silver can introduce discontinuities in the acrylic matrix, reducing its durability, but in certain quantities the silver is also able to improve the protective performance of the composite coating. From an aesthetic point of view, colloidal silver also acts as a particular pigment, the effect of which, however, is significantly influenced by the oxidative and dissolutive phenomena that occur due to exposure to solar radiation.

Thus, colloidal silver represents a suitable filler in cataphoretic processes, but needs a targeted control on the adequate amount, in order to avoid being negative for the durability of the coating itself.

Author Contributions: Conceptualization, M.C. and S.R.; methodology, M.C. and S.R.; investigation, M.C.; data curation, M.C. and S.R.; writing—original draft preparation, M.C.; writing—review and editing, M.C. and S.R.; supervision, S.R. All authors have read and agreed to the published version of the manuscript.

Funding: This research was funded by the RIVID project granted by the University of Trento in 2020 in the context of the "Covid19" call awarded to S.R. (coordinator).

Institutional Review Board Statement: Not applicable.

Informed Consent Statement: Not applicable.

Data Availability Statement: The data presented in this study are available on request from the corresponding author. The data are not publicly available due to the absence of an institutional repository.

Acknowledgments: The authors greatly acknowledge the contributions of Gianluca Gazzola and Marzio Marchesi (Arsonsisi, Lainate MI, Italy) regarding the paint bath supply.

Conflicts of Interest: The authors declare no conflict of interest.

References

1. Abdellah, A.; Fabel, B.; Lugli, P.; Scarpa, G. Spray deposition of organic semiconducting thin-films: Towards the fabrication of arbitrary shaped organic electronic devices. *Org. Electron.* **2010**, *11*, 1031–1038. [CrossRef]
2. Aleksandrova, M.; Andreev, S.; Kolev, G. Spray deposition of organic electroluminescent coatings for application in flexible light emitting devices. *Cogent Eng.* **2015**, *2*, 1014248. [CrossRef]
3. Girotto, C.; Rand, B.P.; Genoe, J.; Heremans, P. Exploring spray coating as a deposition technique for the fabrication of solution-processed solar cells. *Sol. Energy Mater. Sol. Cells* **2009**, *93*, 454–458. [CrossRef]
4. Sparks, B.J.; Hoff, E.F.T.; Xiong, L.; Goetz, J.T.; Patton, D.L. Superhydrophobic Hybrid inorganic–organic thiol-ene surfaces fabricated via spray-deposition and photopolymerization. *ACS Appl. Mater. Interfaces* **2013**, *5*, 1811–1817. [CrossRef] [PubMed]
5. Ramdé, T.; Ecco, L.G.; Rossi, S. Visual appearance durability as function of natural and accelerated ageing of electrophoretic styrene-acrylic coatings: Influence of yellow pigment concentration. *Prog. Org. Coat.* **2017**, *103*, 23–32. [CrossRef]
6. Deflorian, F.; Rossi, S.; Prosseda, S. Improvement of corrosion protection system for aluminium body bus used in public transportation. *Mater. Des.* **2006**, *27*, 758–769. [CrossRef]
7. Rossi, S.; Calovi, M.; Fedel, M. Corrosion protection of aluminum foams by cataphoretic deposition of organic coatings. *Prog. Org. Coat.* **2017**, *109*, 144–151. [CrossRef]
8. Karacal, P.N.; Elginoz, N.; Babuna, F.G. Environmental burdens of cataphoresis process. *Desalin. W. Treat.* **2019**, *172*, 301–308. [CrossRef]
9. Almeida, E.; Alves, I.; Brites, C.; Fedrizzi, L. Cataphoretic and autophoretic automotive primers: A comparative study. *Prog. Org. Coat.* **2003**, *46*, 8–20. [CrossRef]
10. Skotnicki, W.; Jędrzejczyk, D. The comparative analysis of the coatings deposited on the automotive parts by the cataphoresis method. *Materials* **2021**, *14*, 6155. [CrossRef]
11. Youssouf, H.; Azzouzi, S.R.; Reklaoui, K. Control of the "Cataphoresis" Process by Artificial Intelligence. In Proceedings of the Advances in Intelligent Systems and Computing, Marrakech, Morocco, 8–11 July 2019; pp. 169–188.
12. Hylák, K.; Matuška, Z.; Drašnar, P.; Kudláček, J.; Horník, J. Development of equipment for mass cataphoresis painting. *Mater. Sci. Forum* **2019**, *952*, 92–98. [CrossRef]
13. Benato, A. Improving the efficiency of a cataphoresis oven with a cogenerative organic Rankine cycle unit. *Therm. Sci. Eng. Prog.* **2018**, *5*, 182–194. [CrossRef]
14. García, S.; Rodríguez, M.T.; Izquierdo, R.; Suay, J. Evaluation of cure temperature effects in cataphoretic automotive primers by electrochemical techniques. *Prog. Org. Coat.* **2007**, *60*, 303–311. [CrossRef]
15. García, S.; Suay, J. Optimization of deposition voltage of cataphoretic automotive primers assessed by EIS and AC/DC/AC. *Prog. Org. Coat.* **2009**, *66*, 306–313. [CrossRef]
16. Araújo, E.; Rodrigues, M.A.; Viana, A.; Viana, R. The influence of glycerol as an additive in Zinc-Manganese alloy coatings formed by electrodeposition. *Acta Sci. Technol.* **2019**, *41*, 41103. [CrossRef]
17. Bučko, M.; Mišković-Stanković, V.; Rogan, J.; Bajat, J.B. The protective properties of epoxy coating electrodeposited on Zn-Mn alloy substrate. *Prog. Org. Coat.* **2015**, *79*, 8–16. [CrossRef]
18. Szklarska, M.; Łosiewicz, B.; Dercz, G.; Maszybrocka, J.; Rams-Baron, M.; Stach, S. Electrophoretic deposition of chitosan coatings on the Ti15Mo biomedical alloy from a citric acid solution. *RSC Adv.* **2020**, *10*, 13386–13393. [CrossRef]
19. Zanella, C.; Pedrotti, A.; Fedel, M.; Deflorian, F. Influence of the electrochemical behavior of metal substrates on the properties of cataphoretic clearcoat. *Prog. Org. Coat.* **2014**, *77*, 1987–1992. [CrossRef]
20. Živković, L.S.; Bajat, J.B.; Popić, J.P.; Jegdić, B.V.; Stevanović, S.; Mišković-Stanković, V.B. Protective properties of cataphoretic epoxy coating on aluminium alloy AA6060 modified with electrodeposited Ce-based coatings: Effect of post-treatment. *Prog. Org. Coat.* **2015**, *79*, 43–52. [CrossRef]
21. Romano, A.P.; Olivier, M.G.; Nazarov, A.; Thierry, D. Influence of crosslinking density of a cataphoretic coating on initiation and propagation of filiform corrosion of AA6016. *Prog. Org. Coat.* **2009**, *66*, 173–182. [CrossRef]
22. Reichinger, M.; Bremser, W.; Dornbusch, M. Interface and volume transport on technical cataphoretic painting: A comparison of steel, hot-dip galvanised steel and aluminium alloy. *Electrochim. Acta* **2017**, *231*, 135–152. [CrossRef]
23. Romano, A.-P.; Fedel, M.; Deflorian, F.; Olivier, M.G. Silane sol–gel film as pretreatment for improvement of barrier properties and filiform corrosion resistance of 6016 aluminium alloy covered by cataphoretic coating. *Prog. Org. Coat.* **2011**, *72*, 695–702. [CrossRef]
24. Fedel, M. Effect of sol–gel layers obtained from GLYMO/MTES mixtures on the delamination of a cataphoretic paint on AA1050. *J. Coat. Technol. Res.* **2017**, *14*, 425–435. [CrossRef]
25. Zivkovic, L.S.; Jegdjc, B.V.; Andric, V.; Rhee, K.Y.; Bajat, J.B.; Miakovjc Stankovic, V. The effect of ceria and zirconia nanoparticles on the corrosion behaviour of cataphoretic epoxy coatings on AA6060 alloy. *Prog. Org. Coat.* **2019**, *136*, 105219. [CrossRef]
26. Jegdić, B.V.; Živković, L.S.; Popić, J.P.; Rogan, J.; Bajat, J.B.; Mišković-Stanković, V.B. Corrosion stability of cerium-doped cataphoretic epoxy coatings on AA6060 alloy. *Mater. Corros.* **2016**, *67*, 1173–1184. [CrossRef]
27. Rossi, S.; Calovi, M. Addition of graphene oxide plates in cataphoretic deposited organic coatings. *Prog. Org. Coat.* **2018**, *125*, 40–47. [CrossRef]
28. Calovi, M.; Dirè, S.; Ceccato, R.; Deflorian, F.; Rossi, S. Corrosion protection properties of functionalised graphene–acrylate coatings produced via cataphoretic deposition. *Prog. Org. Coat.* **2019**, *136*, 105261. [CrossRef]

29. Calovi, M.; Rossi, S.; Deflorian, F.; Dirè, S.; Ceccato, R. Effect of functionalized graphene oxide concentration on the corrosion resistance properties provided by cataphoretic acrylic coatings. *Mater. Chem. Phys.* **2019**, *239*, 121984. [CrossRef]
30. Calovi, M.; Rossi, S.; Deflorian, F.; Dirè, S.; Ceccato, R.; Guo, X.; Frankel, G.S. Effects of graphene-based fillers on cathodic delamination and abrasion resistance of cataphoretic organic coatings. *Coatings* **2020**, *10*, 602. [CrossRef]
31. Calovi, M.; Russo, F.; Rossi, S. Synergic behavior of graphene-based filler and thermochromic pigments in cataphoretic coatings. *Prog. Org. Coat.* **2021**, *150*, 105978. [CrossRef]
32. Calovi, M.; Russo, F.; Rossi, S. Esthetic performance of thermochromic pigments in cataphoretic and sprayed coatings for outdoor applications. *J. Appl. Polym. Sci.* **2021**, *138*, 50622. [CrossRef]
33. El-Faham, A.; Atta, A.M.; Osman, S.M.; Ezzat, A.O.; El-saeed, A.M.; Al Othman, Z.A.; Al-Lohedan, H.A. Silver-embedded epoxy nanocomposites as organic coatings for steel. *Prog. Org. Coat.* **2018**, *123*, 209–222. [CrossRef]
34. Giraldo Mejía, H.F.; Herrera Seitz, K.; Valdés, M.; Uheida, A.; Procaccini, R.A.; Pellice, S.A. Antibacterial performance of hybrid nanocomposite coatings containing clay and silver nanoparticles. *Colloids Surf. A Physicochem. Eng. Asp.* **2021**, *628*, 127354. [CrossRef]
35. Rozilah, A.; Aiza Jaafar, C.N.; Sapuan, S.M.; Zainol, I.; Ilyas, R.A. The effects of silver nanoparticles compositions on the mechanical, physiochemical, antibacterial, and morphology properties of sugar palm starch biocomposites for antibacterial coating. *Polymers* **2020**, *12*, 2605. [CrossRef]
36. Rai, N.K.; Ashok, A.; Akondi, B.R. Consequences of chemical impact of disinfectants: Safe preventive measures against COVID-19. *Crit. Rev. Toxicol.* **2020**, *50*, 513–520. [CrossRef] [PubMed]
37. Russell, A.; Hugo, W. 7 antimicrobial activity and action of silver. *Prog. Med. Chem.* **1994**, *31*, 351–370. [PubMed]
38. Karunagaran, V.; Rajendran, K.; Sen, S. Antimicrobial activity of biosynthesized silver oxide nanoparticles. *J. Pure Appl. Microbiol.* **2014**, *4*, 3263–3268.
39. Swathy, J.R.; Sankar, M.U.; Chaudhary, A.; Aigal, S.; Anshup; Pradeep, T. Antimicrobial silver: An unprecedented anion effect. *Sci. Rep.* **2014**, *4*, 7161. [CrossRef]
40. Schneider, G. Antimicrobial silver nanoparticles–Regulatory situation in the European Union. *Mater. Today Proc.* **2017**, *4*, S200–S207. [CrossRef]
41. Yin, I.; Zhang, J.; Zhao, I.; Mei, M.L.; Li, Q.; Chu, C.-H. The Antibacterial mechanism of silver nanoparticles and its application in dentistry. *Int. J. Nanomed.* **2020**, *15*, 2555–2562. [CrossRef]
42. Vimbela, G.; Sang, N.; Fraze, C.; Yang, L.; Stout, D. Antibacterial properties and toxicity from metallic nanomaterials [Corrigendum]. *Int. J. Nanomed.* **2018**, *13*, 6497–6498. [CrossRef] [PubMed]
43. Noronha, V.T.; Paula, A.J.; Durán, G.; Galembeck, A.; Cogo-Müller, K.; Franz-Montan, M.; Duran, N. Silver nanoparticles in dentistry. *Dent. Mater.* **2017**, *33*, 1110–1126. [CrossRef] [PubMed]
44. Gonzalo-Juan, I.; Xie, F.; Becker, M.; Tulyaganov, D.U.; Ionescu, E.; Lauterbach, S.; De Angelis Rigotti, F.; Fischer, A.; Riedel, R. Synthesis of silver modified bioactive glassy materials with antibacterial properties via facile and low-temperature route. *Materials* **2020**, *13*, 5115. [CrossRef] [PubMed]
45. da Silva, L.C.A.; Neto, F.G.; Pimentel, S.S.C.; Palácios, R.d.S.; Sato, F.; Retamiro, K.M.; Fernandes, N.S.; Nakamura, C.V.; Pedrochi, F.; Steimacher, A. The role of Ag2O on antibacterial and bioactive properties of borate glasses. *J. Non-Cryst. Sol.* **2021**, *554*, 120611. [CrossRef]
46. Dharmaraj, D.; Krishnamoorthy, M.; Rajendran, K.; Karuppiah, K.; Annamalai, J.; Durairaj, K.R.; Santhiyagu, P.; Ethiraj, K. Antibacterial and cytotoxicity activities of biosynthesized silver oxide (Ag2O) nanoparticles using Bacillus paramycoides. *J. Drug Deliv. Sci. Technol.* **2021**, *61*, 102111. [CrossRef]
47. Agbe, H.; Sarkar, D.K.; Chen, X.G. Electrochemically synthesized silver phosphate coating on anodized aluminum with superior antibacterial properties. *Surf. Coat. Technol.* **2021**, *428*, 127892. [CrossRef]
48. Calovi, M.; Furlan, B.; Coroneo, V.; Massidda, O.; Rossi, S. Facile route to effective antimicrobial aluminum oxide layer realized by co-deposition with silver nitrate. *Coatings* **2022**, *12*, 28. [CrossRef]
49. Dehghan, F.; Mardanpour, H.; Kamali, S.; Alirezaei, S. Synthesis and antibacterial properties of novel Al_2O_3-Ag anodised composite coating. *Mater. Technol.* **2021**, *36*, 721–730. [CrossRef]
50. Zhang, L.; Li, B.; Zhang, X.; Wang, D.; Zhou, L.; Li, H.; Liang, C.; Liu, S.; Wang, H. Biological and antibacterial properties of TiO_2 coatings containing Ca/P/Ag by one-step and two-step methods. *Biomed. Microdevices* **2020**, *22*, 24. [CrossRef]
51. Chen, D.; Li, Y.; He, H.; Li, W.; Zeng, R.; Wang, X. Covalent incorporation of Ag nanoparticles into TiO_2 nanotubes on Ti_6Al_4V by molecular grafting for enhancing antibacterial effect. *Surf. Coat. Technol.* **2021**, *426*, 127773. [CrossRef]
52. Makowski, T.; Svyntkivska, M.; Piorkowska, E.; Mizerska, U.; Fortuniak, W.; Kowalczyk, D.; Brzezinski, S.; Kregiel, D. Antibacterial electroconductive composite coating of cotton fabric. *Materials* **2022**, *15*, 1072. [CrossRef] [PubMed]
53. Chen, Y.; Li, W.; Wang, W.; Zhao, Y.; Chen, M. Microstructure, corrosion resistance, and antibacterial properties of an Ag/Mg-Al layered double hydroxide coating synthesized in situ on biomedical Mg-Zn-Ca alloy. *Ceram. Int.* **2022**, *48*, 4172–4187. [CrossRef]
54. Ganguli, P.; Chaudhuri, S. Nanomaterials in antimicrobial paints and coatings to prevent biodegradation of man-made surfaces: A review. *Mater. Today Proc.* **2020**, *45*, 3769–3777. [CrossRef]
55. Abdullayev, E.; Sakakibara, K.; Okamoto, K.; Wei, W.; Ariga, K.; Lvov, Y. Natural tubule clay template synthesis of silver nanorods for antibacterial composite coating. *ACS Appl. Mater. Interfaces* **2011**, *3*, 4040–4046. [CrossRef] [PubMed]

56. Cheng, L.; Ren, S.; Lu, X. Application of eco-friendlywaterborne polyurethane composite coating incorporated with nano cellulose crystalline and silver nano particles on wood antibacterial board. *Polymers* **2020**, *12*, 407. [CrossRef] [PubMed]
57. Yan, X.; Pan, P. Preparation of silver antibacterial agents with different forms and their effects on the properties of water-based primer on tilia europaea surface. *Coatings* **2021**, *11*, 1066. [CrossRef]
58. Wang, L.; Porto, C.L.; Palumbo, F.; Modic, M.; Cvelbar, U.; Ghobeira, R.; De Geyter, N.; De Vrieze, M.; Serša, G.; Leys, C.; et al. Synthesis of antibacterial composite coating containing nanocapsules in an atmospheric pressure plasma. *Mater. Sci. Eng. C* **2021**, *119*, 111496. [CrossRef]
59. Lateef, A.; Azeez, M.A.; Asafa, T.B.; Yekeen, T.A.; Akinboro, A.; Oladipo, I.C.; Azeez, L.; Ajibade, S.E.; Ojo, S.A.; Gueguim-Kana, E.B.; et al. Biogenic synthesis of silver nanoparticles using a pod extract of Cola nitida: Antibacterial and antioxidant activities and application as a paint additive. *J. Taibah Univ. Sci.* **2016**, *10*, 551–562. [CrossRef]
60. Deya, C.; Bellotti, N. Biosynthesized silver nanoparticles to control fungal infections in indoor environments. *Adv. Nat.Sci. Nanosci. Nanotechnol.* **2017**, *8*, 025005. [CrossRef]
61. Barberia-Roque, L.; Gámez-Espinosa, E.; Viera, M.; Bellotti, N. Assessment of three plant extracts to obtain silver nanoparticles as alternative additives to control biodeterioration of coatings. *Int. Biodeterior. Biodegrad.* **2019**, *141*, 52–61. [CrossRef]
62. Liu, S.; He, J.; Xue, J.; Ding, W. Efficient fabrication of transparent antimicrobial poly(vinyl alcohol) thin films. *J. Nanopart. Res.* **2009**, *11*, 553–560. [CrossRef]
63. Bechtold, M.; Valério, A.; Souza, A.; Oliveira, D.; Franco, C.; Serafim, R.; Guelli Souza, S. Synthesis and application of silver nanoparticles as biocidal agent in polyurethane coating. *J. Coat. Technol. Res.* **2020**, *17*, 613–620. [CrossRef]
64. Asafa, T.; Odediji, R.; Salaudeen, T.; Lateef, A.; Durowoju, M.; Azeez, M.; Yekeen, T.; Oladipo, I.; Irshad, H.M.; Hakeem, A. Physico-mechanical properties of emulsion paint embedded with silver nanoparticles. *Bull. Mater. Sci.* **2021**, *44*, 7. [CrossRef]
65. Ohashi, F.; Shibahara, A. Antibacterial and antifungal properties of clear coating film containing silver–cytokinin complex as a filler. *J. Coat. Technol. Res.* **2020**, *17*, 1619–1623. [CrossRef]
66. *ASTM B117:2011*; Operating Salt Spray (Fog) Apparatus. ASTM: West Conshohocken, PA, USA, 2011; pp. 1–12.
67. *ASTM G154:16*; Standard Practice for Operating Fluorescent Ultraviolet (UV) Lamp Apparatus for Exposure of Nonmetallic Materials. ASTM: West Conshohocken, PA, USA, 2016.
68. *ISO 4628*; Evaluation of Degradation of Coatings. ISO: Geneva, Switzerland, 2012.
69. Deflorian, F.; Rossi, S. An EIS study of ion diffusion through organic coatings. *Electrochim. Acta* **2006**, *51*, 1736–1744. [CrossRef]
70. Amirudin, A.; Thieny, D. Application of electrochemical impedance spectroscopy to study the degradation of polymer-coated metals. *Prog. Org. Coat.* **1995**, *26*, 1–28. [CrossRef]
71. Akbarinezhad, E.; Bahremandi, M.; Faridi, H.; Rezaei, F. Another approach for ranking and evaluating organic paint coatings via electrochemical impedance spectroscopy. *Corros. Sci.* **2009**, *51*, 356–363. [CrossRef]
72. Ghazizadeh, A.; Haddadi, S.A.; Mahdavian, M. The effect of sol–gel surface modified silver nanoparticles on the protective properties of the epoxy coating. *RSC Adv.* **2016**, *6*, 18996–19006. [CrossRef]
73. Iannucci, L.; Parvis, M.; Sangermano, M.; Angelini, E.; Grassini, S. Rivestimenti ibridi organico-inorganici per la protezione dalla corrosione. *La Metall. Ital.* **2020**, *11*, 60–63.
74. Procaccini, R.; Bouchet, A.; Pastore, J.I.; Studdert, C.; Ceré, S.; Pellice, S. Silver-functionalized methyl-silica hybrid materials as antibacterial coatings on surgical-grade stainless steel. *Prog. Org. Coat.* **2016**, *97*, 28–36. [CrossRef]
75. Anghelone, M.; Jembrih-Simbürger, D.; Schreiner, M. Influence of phthalocyanine pigments on the photo-degradation of alkyd artists' paints under different conditions of artificial solar radiation. *Polym. Degrad. Stab.* **2016**, *134*, 157–168. [CrossRef]
76. Gulmine, J.; Janissek, P.; Heise, H.; Akcelrud, L. Degradation profile of polyethylene after artificial accelerated weathering. *Polym. Degrad. Stab.* **2003**, *79*, 385–397. [CrossRef]
77. Chiantore, O.; Trossarelli, L.; Lazzari, M. Photooxidative degradation of acrylic and methacrylic polymers. *Polymer* **2000**, *41*, 1657–1668. [CrossRef]
78. Kaczmarek, H.; Kamińska, A.; van Herk, A. Photooxidative degradation of poly (alkyl methacrylate) s. *Eur. Polym. J.* **2000**, *36*, 767–777. [CrossRef]
79. Mittelman, A.M.; Fortner, J.D.; Pennell, K.D. Effects of ultraviolet light on silver nanoparticle mobility and dissolution. *Environ. Sci. Nano* **2015**, *2*, 683–691. [CrossRef]
80. Odzak, N.; Kistler, D.; Sigg, L. Influence of daylight on the fate of silver and zinc oxide nanoparticles in natural aquatic environments. *Environ. Pollut.* **2017**, *226*, 1–11. [CrossRef]
81. Lin, C.-C.; Lin, D.-X.; Lin, S.-H. Degradation problem in silver nanowire transparent electrodes caused by ultraviolet exposure. *Nanotechnology* **2020**, *31*, 215705. [CrossRef]
82. *ASTM E308-18*; Standard Practice for Computing the Colors of Objectives by Using the CIE System. ASTM: West Conshohocken, PA, USA, 2018.
83. Hasani, M.; Mahdavian, M.; Yari, H.; Ramezanzadeh, B. Versatile protection of exterior coatings by the aid of graphene oxide nano-sheets; comparison with conventional UV absorbers. *Prog. Org. Coat.* **2018**, *116*, 90–101. [CrossRef]
84. Herley, P.J.; Prout, E. The thermal decomposition of silver oxide. *J. Am. Chem. Soc.* **1960**, *82*, 1540–1543. [CrossRef]

coatings

MDPI

Article

Study of the Durability and Aesthetical Properties of Powder Coatings Admixed with Pearlescent Pigments

Stefano Rossi *, Francesca Russo and Lotfi Bouchakour Rahmani

Department of Industrial Engineering, University of Trento, Via Sommarive 9, 38123 Trento, Italy; francesca.russo-2@unitn.it (F.R.); cjls@libero.it (L.B.R.)

* Correspondence: stefano.rossi@unitn.it; Tel.: +39-0461-282442

Received: 30 January 2020; Accepted: 28 February 2020; Published: 3 March 2020

Abstract: Powder coatings are commonly used to protect metallic substrates because of their good protection properties together with their aesthetical ones. In this work, the aesthetic and functional properties of powder coatings admixed with pearlescent pigments were investigated. The main aim of this work is represented by the evaluation of the coating properties and their change assessment after accelerated ageing treatments, such as UV and salt spray exposure. Changes in gloss, color and roughness were recorded before, during, and after UVA and UVB exposure. All the samples showed good resistance to UVA irradiation, whereas UVB light caused a huge variation in surface properties. Further insights on the degradation of the polymeric matrix were gained by exploiting FT-IR analyses, which could be correlated to the change of color. The protection properties of the coating were assessed by exploiting an acetic salt spray test. All the samples showed good resistance to aggressive environments, and no pigment effect on the degradation of the coatings could be detected.

Keywords: pearlescent pigments; powder coatings; durability; ageing treatments

1. Introduction

Nowadays, the main challenge of materials engineering is the modification of the surface characteristics of a substrate in order to improve its functional properties for high duty technological applications. In the last decades, the development of a coating has increasingly focused on finding materials with surprising functional properties that also had good perceptive and aesthetical characteristics [1]. In this context, organic coatings are employed to guarantee the good protection of metal products, combining a high sustainability of the industrial deposition process [1] with the possibility of obtaining a high variety of surface aesthetic effects. Thus, the evaluation of the aesthetical properties is a key factor to take under consideration when studying the coating performances [2]. Among all the different types of protective layers, powder coatings are arousing increasing interest in the building, automotive, and cosmetic industry, mainly due to the low environmental and economic impact they entail and for the wide range of pigments available on the market [3]. Pigments consist of insoluble powders dispersed in the coating matrix, and they can be either organic or inorganic, though the latest type is the most used [4]. To understand the industrial importance of pigment production, it is sufficient to report that in 2005 the world production of inorganic pigments was approximately equal to 6 million tons, of which at least half was for the production of high-performance pigments [5]. The industrial research on pigments has recently focused on three main roads: the economization of already existing pigments, the discovery and development of inorganic pigments with better performance features, and the replacement of some toxic and non-ecological chemical substances in order to be in compliance with new national and international regulations [6,7].

Pigments can be classified in different ways, according, for example, to the way they interact with light; among them, this work will focus on the special effects pigments, and in particular on pearlescent pigments [8]. Pearlescent pigments give the coating additional color effects, such as an angular color dependence, which is mainly due to the interference of light on thin layers, flakes, or platelets [9]. The industrial production of pearlescent pigments started during the 1920s using mercury and arsenic salts. In 1963, the discovery of new TiO_2-coated mica pigments was reported, and only in the 1990s were effect pigments, based on aluminum platelets and coated with ferric oxide, introduced into the market [10].

Effect pigments can be basically divided into two types: substrate-free pigments and layered pigments. The first show a brittle behavior, and their use is limited by their chemical composition, whereas the latter are produced by coating thin laminar platelets of mica or alumina with high refractive optical layers, such as titanium dioxide, iron oxide, or a combination of both [11]. Most pearlescent pigments have at least three layers of two different materials with different refractive indexes. The pearlescent effect is produced by the specular reflection of light from the many surfaces of the platelets at various depths within the coating film. Light striking the platelet is partially reflected and partially transmitted through the platelet; the pearlescent effect is produced by the dependence of the reflection on the viewing angle [12]. Alumina-based pigments have a strong pearlescent effect with respect to mica-based pigments because they are easily produced, with a very narrow thickness distribution and very smooth surfaces [11]. In the last decade, this type of pigments has found a broad application range, both for decorative and functional purposes [13,14], also thanks to the ease of incorporation in the coating matrix [11]. Alumina-based pigments exhibit the well-known advantages of mica pigments together with the possibility of realizing unique optical effects thanks to their controlled thickness and chemical purity [11].

This study is focused on the characterization of powder coatings with the addition of innovative alumina-based pearlescent pigments. This type of pigments has got enhanced optical properties and shows great pearlescent behavior with respect to mica-based pigments, but no studies about their effect on the durability of the coatings can be found in the literature. The main aim of this work is represented by the evaluation of coatings' aesthetical and functional properties and their change assessment as a consequence of accelerated ageing treatments, such as UV and salt spray exposure.

2. Materials and Methods

The superdurable powder coatings Interpon D2525, based on isophthalic acid (IPA) polyester resin and TGIC-free, were supplied by Akzo Nobel Coatings S.p.A. (Como, Italy) and were applied on $150 \times 60 \times 0.75$ mm^3 5005 aluminum alloy panels. Regarding the coating formulation, five different pigment types, supplied by Merck S.p.A (Darmstadt, Germany), were added to the polymeric matrix, thus obtaining the five sample series reported in Table 1. In the premixing phase, a special treatment, consisting in raising the temperature by a few degrees above the matrix glass transition temperature (T_g), was exploited in order to obtain a very homogeneous powder, allowing the pigments to be uniformly dispersed. Therefore, the layers were applied by an electrostatic spray deposition method and cured at 200 °C for 15 min.

Table 1. Nomenclature of the samples with the component formulation.

Coating Name	Pigment Type	Pigment wt.%	Powder Coating wt.%
GB-SW	Xirallic T60-23 SW Galaxy Blue	4.00	96.00
GB-WNT	Xirallic T60-23 WNT Galaxy Blue	4.00	96.00
LG	Xirallic NXT M260-30 SW Leonis Gold	4.00	96.00
PS	Xirallic NXT M60-69 WNT Panthera Silver	4.00	96.00
TB	Xirallic NXT M260-23 Tigris Blue	4.00	96.00

The pigments color and morphology were studied by optical and electron microscopy, by means of Nikon SMZ25 stereomicroscope (Nikon Instruments, Amstelveen, Netherlands) and JEOL IT300 Scanning Electron Microscope (SEM, Jeol Ltd., Akishima, Tokyo, Japan) microscope, respectively. Moreover, SEM observations were employed in the coatings and pigments thickness assessment. The chemical composition of the pigments was analyzed by Energy Dispersive X-Ray Spectrometry (EDS), carried out using a Bruker "Quantax Micro-XRF" analyzer (Billerica, MA, USA). The influence of the environment on the aesthetical properties of the coatings was assessed by exposing the five sample series to UV radiation (UVA and UVB radiation for times up to 1000 h). UV radiation is responsible for the degradation processes of organic coatings, since the associated energy is sufficient to break the C–H and C–C bonds of the polymer. UVA and UVB radiations are distinguished by the wavelength range, which ranges from 400 to 320 nm for UVA light and from 320 to 280 nm for UVB light. UV light accelerated tests were carried out using a UVA-340 lamp and a UVB-313EL lamp with emission peaks at 340 and 313 nm, respectively, as described by the ASTM G154 (2016) standard [15], in order to check how different radiation types affected the degradation of the coating. The changes of aesthetical properties were also evaluated by exploiting color measurements, using the spectrophotometer CM-2600d Konica Minolta, with a D65/10° illuminant/observer, and also by exploiting gloss and roughness measurements, according to the ASTM D523-14 (2018) [16] standard and ISO 4288 (1996) standard [17], respectively. The gloss measurements were collected using an Erichsen NL3A glossmeter (Westlake, OH, USA), acquiring the values for the residual gloss at 60°. Every reported value represents the average of six measurements. The roughness data were acquired by means of the MAHR Marsurf PS1 roughness tester (Esslingen, Germany). A sampling length (l_r) of 0.8 mm and an evaluation length (l_n) of 5.6 mm was used; the reported value is the average of five measurements. In order to check the deterioration of the coating organic matrix, IR spectra were collected between 4000 and 600 cm^{-1} before and after UVA and UVB exposure, using the spectrophotometer Varian 4100 FT-IR with a resolution of 4 cm^{-1} (Palo Alto, CA, USA). Regarding the protection properties of the coating, accelerated corrosion testing was exploited in an acetic acid salt spray chamber on as-made samples and also on samples with the presence of an artificial notch. The samples were exposed for times up to 1000 h to a salt spray solution produced by a 5 wt.% NaCl solution at 35 °C and 100% RH, according to the ASTM B117-19 (2019) standard [18], with the addition of acetic acid in order to vary the pH between 3.1 and 3.3, thus avoiding the passivation of Al substrate at neutral pH, as described by the ASTM G85-19 (2019) standard [19]. As a first thing, the salt spray test was involved in the study of the blister evolution and detachment grade of the coating, by making a cross-cut incision of 1 mm width down to the metal, as specified by the UNI EN ISO 4628-8 (2013) [20]. In addition to that, Electrochemical Impedance Spectroscopy (EIS) measurements were carried out on as-made samples after 0, 400, 750, and 1000 h of exposition in a salt spray chamber, in order to evaluate the protection properties of the coatings. The measurements have been exploited in 5 wt.% NaCl solution and using a three-electrode configuration: a platinum electrode was used as a counter electrode, and an Ag/AgCl electrode was used as a reference. The tested area was 4.9 cm^2. A signal amplitude of 30 mV in the frequency range from 10^6 to 10^{-2} Hz was set up.

3. Results

3.1. Morphology and Chemical Characterization

Figure 1 shows the morphology of the pigment flakes. The pigments are free-flowing powders, and their average diameter dimension is included between 10 and 30 μm. From a detailed SEM analysis at high magnifications, it is also possible to assess the pigments' thickness, which resulted as being about 550, 590, 660, 250, and 460 nm for the GB-SW, GB-WNT, LG, PS, and TB pigments, respectively. The difference in thickness among the pigments, which is lower than 1 μm in all cases, can also be noticed from their different transparency to electrons, in particular regarding the PS and TB pigments (Figure 1).

Figure 1. SEM images of pigment flakes: (**a**) GB-SW; (**b**) GB-WNT; (**c**) LG; (**d**) PS; and (**e**) TB.

The chemical composition of the pigments was investigated through a semiquantitative EDS analysis, whose results in atomic percentages are reported in Table 2. The results are extrapolated by a fitting operated by the Esprit Bruker software (2.0).

Table 2. EDS analysis of the studied pigments.

Pigment	O %	Si %	Al %	Ti %	Sn %	Zr %	Fe%	Cu %
Xirallic T60-23 SW Galaxy Blue	51.0	1.4	28.1	18.1	1.4	–	–	–
Xirallic T60-23 WNT Galaxy Blue	48.4	1.3	31.4	18.1	0.3	0.3	0.1	0.1
Xirallic NXT M260-30 SW Leonis Gold	40.6	0.8	29.1	10.7	1.0	–	17.8	–
Xirallic NXT M60-69 WNT Phantera Silver	49.2	0.5	41.5	5.9	0.2	0.6	2.1	–
Xirallic NXT M260-23 Tigris Blue	52.1	1.4	28.4	17.8	0.3	–	–	–

Comparing the five different pigment types, it is possible to detect for all of them a high presence of aluminum and titanium: aluminum, present as α-Al_2O_3, is the main component of the pigment substrate, whereas titanium, deposited as TiO_2 on the substrate, is used as an optical layer, together with other oxides, present at lower concentrations [11]. All the samples show the presence of silicon and tin, used to promote the nucleation of the rutile phase of TiO_2. The presence of the rutile phase is to be preferred to the anatase one, because the latter has a lower photocatalytic activity. Several studies have reported on the investigation of the effect of different elements on the rutile-anatase phase transformation in pearlescent pigments: the presence of tin oxide, in the crystalline form of casserite, should promote the epitaxial growth of the rutile phase, thanks to the perfect match between their cell parameters [21,22]. The Leonis Gold pigment contains a very high amount of Fe, also present as an oxide, to which the intense yellow color of the pigment and the high light interference effect are attributable. Only the Galaxy Blue WNT and Panthera Silver pigments show the presence of zirconium oxide, which is a high-refractive material and which is in widespread use for the synthesis of pearlescent pigments [23].

The SEM micrographs in Figure 2 show the as-deposited sample cross sections. All the sample cross-sections highlight a good and homogeneous dispersion of the pigments inside the polymeric matrix. The random orientation of the pigments, as shown in Figure 2, has a positive effect in harvesting the coating pearlescent effect. The thickness of the samples can be also evaluated from the SEM micrographs in Figure 2, and its value is between 75 and 150 μm. The high thickness dispersion of the coating is typical of powder coatings obtained by spray deposition and does not affect the properties under investigation.

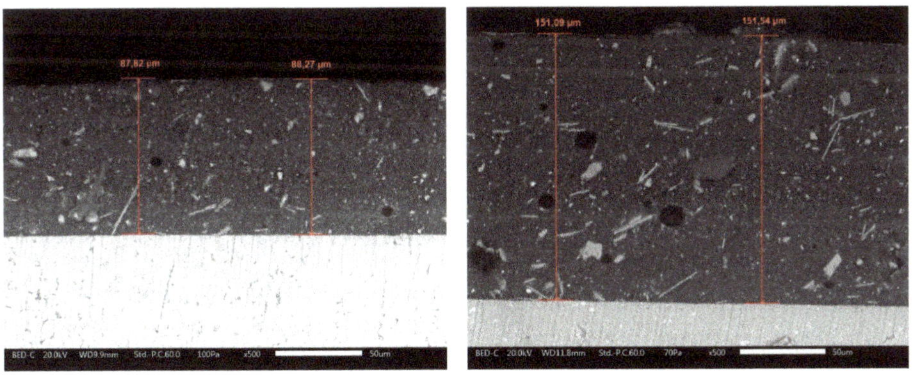

(a) (b)

Figure 2. *Cont.*

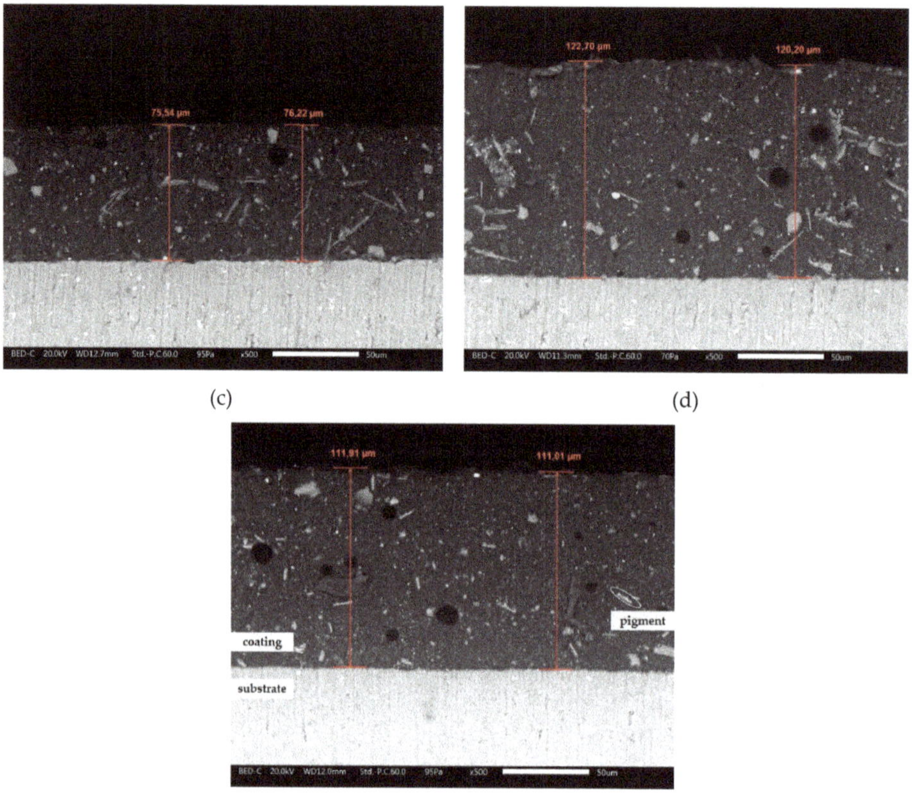

(c) (d)

(e)

Figure 2. Scanning electron microscope cross sections of the sample: (**a**) GB-SW; (**b**) GB-WNT; (**c**) LG; (**d**) PS; and (**e**) TB. In (**e**), a description of the components of the sample is made.

3.2. Gloss and Roughness Evaluation after UV Exposure

The changes of the coating aesthetical properties were assessed after 1000 h UV exposure. Regarding the roughness change, the mean roughness parameter, R_a, was chosen in order to give the average information about the sample surface.

Table 3 shows the roughness variation for all the samples before and after UVA and UVB exposure. All the samples underwent an increase in roughness of about 0.1–0.2 µm after 1000 h of irradiation with UVB light, whereas a different behaviour after 1000 h UVA irradiation was verifiable. All the samples showed a decreased superficial roughness after UVA irradiation, apart from the LG sample, which showed a small increase. Considering the error bars, it is possible to state that the roughness values after UVA exposure could be comparable to the values of the pristine samples, and thus no significant changes could be seen after 1000 h of UVA exposure.

Table 3. Sample roughness before and after 1000 h irradiation with UVA and UVB light.

Coating Name	R_a As-Made (μm)	R_a 1000 h UVA (μm)	R_a 1000 h UVB (μm)
GB-SW	0.66 ± 0.08	0.65 ± 0.05	0.77 ± 0.33
GB-WNT	0.86 ± 0.24	0.79 ± 0.12	0.97 ± 0.10
LG	0.65 ± 0.04	0.69 ± 0.08	0.87 ± 0.08
PS	0.74 ± 0.14	0.69 ± 0.08	0.82 ± 0.12
TB	0.66 ± 0.11	0.59 ± 0.08	0.75 ± 0.11

The surface roughness influences the gloss of a surface: on a smooth mirror-like surface, parallel beams of light are reflected in the same certain direction, whereas on an irregular surface, beams of light are diffusely scattered, and the surface appears to be matt. Determining the gloss change is then important to have further insights on the modification of the sample surface properties. After 1000 h of UVA irradiation, the maximum gloss change was about 5%, and it was almost similar for all the samples. It did not follow a linear decreasing trend, but it remained stable over time. Gloss and roughness are closely interrelated, thus the small changes in roughness are connected and in accordance with the little changes in gloss of the samples. In contrast, with respect to what happened to the samples after irradiation with UVA light, it was possible to see a clear decreasing trend of gloss with the increase of the UVB light exposure time, as shown in Figure 3. All the samples showed the same behaviour, which is probably connected to the degradation of the polymeric matrix.

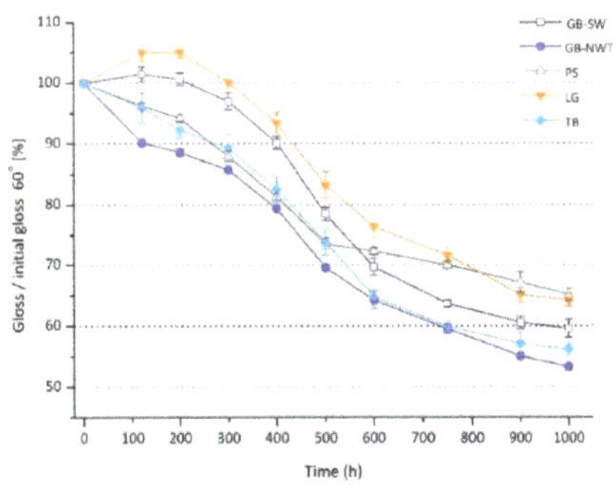

Figure 3. Sample residual gloss after irradiation with UVB light.

Considering the GSB Quality Standard AL 631 [24], which classifies a coating as extremely durable if it maintains at least 50% of the initial gloss after 1000 h UVB irradiation, it is possible to state that all the samples also showed a very good behaviour after 1000 h UVB exposure and met the industrial qualitative standards. The decrease of gloss is closely linked to the increase in roughness, as shown in Table 3.

3.3. Color Evaluation after UV Exposure

Ultraviolet radiation exposure could influence the aesthetic properties of the coatings, changing the colour or aspect of the coating surface. Figure 4a shows the as-prepared samples, and Figure 4b reports their colorimetric coordinates (CIE$L^*a^*b^*$ [25]) in order to appreciate the colour differences among the samples. The colorimetric coordinates values are the average of five measurements at D65/10° in SCE mode; L^* is the lightness (0 is black and 100 is white), a^* is the red-green coordinate (positive values are red, negative values are green), and b^* is the yellow-blue coordinate (positive values are yellow, negative values are blue, and 0 is neutral).

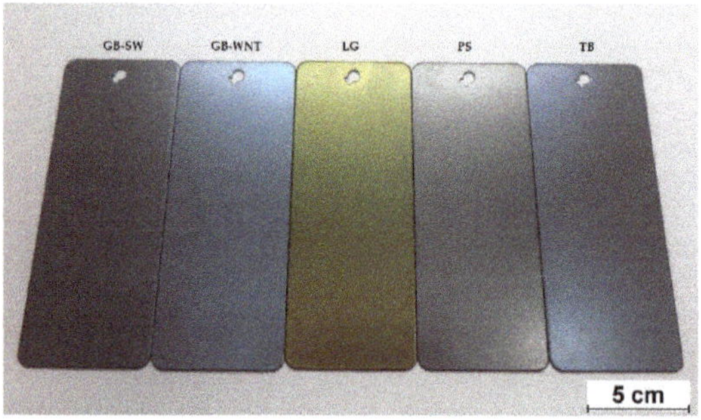

(a)

Sample	L*	a*	b*
GB-SW	52,94 ± 0,05	-0,46 ± 0,01	-0,68 ± 0,10
GB-WNT	53,28 ± 0,02	-0,74 ± 0,02	-1,65 ± 0,09
LG	52,63 ± 0,08	0,27 ± 0,02	9,45 ± 0,16
PS	50,30 ± 0,09	0,13 ± 0,03	-0,17 ± 0,07
TB	53,22 ± 0,14	-0,61 ± 0,01	-1,36 ± 0,10

(b)

Figure 4. (**a**) Photography of the as-prepared samples; (**b**) Samples' colorimetric coordinates according to CIE$L^*a^*b^*$.

All the samples showed a high lightness. Regarding the a^* value, it is possible to see that only the PS and LG samples had positive values, whereas, considering the b^* coordinate, all the samples showed a negative value, apart from the LG sample, which had a high shift towards yellow.

The colour of the coating was evaluated at the same cycles respect to the gloss and roughness measurements. The variation of ΔE, ΔL*, Δa*, Δb* are reported as a function of the exposure time. The symbol ΔE represents the total colour change and it is calculated as follows:

$$\Delta E = [(\Delta L^*)^2 + (\Delta a^*)^2 + (\Delta b^*)^2]^{1/2}, \quad (1)$$

(ASTM E308(2018) [26]). First of all, it is necessary to point out that no significant changes of ΔE, ΔL^*, Δa^*, Δb^* after UVA exposure were recorded. The ΔE variation was lower than 0.40 after 1000 h irradiation, and it did not increase with the exposure time. The variation of the a^* coordinate was negligible, whereas the variation of the b^* coordinate followed a blue shift trend for all the samples and in particular for the PS sample, which had a Δb^* equal to 0.60. Thus, it is possible to state that after irradiation with UVA light the ΔE variation was mainly due to the change of the b^* coordinate and was limited to low values (colour changes with $\Delta E < 1$ are imperceptible to the human eye).

Figure 5 reports the variation of the sample colorimetric coordinates after irradiation with UVB light.

The ΔE variation followed a sigmoidal trend for all the samples, with values after 1000 h irradiation included between 1.4 and 1.9, much higher than the threshold value of $\Delta E = 1$. Despite all the coatings being made of the same polymetric matrix, it was possible to observe the GB-NWT and PS samples having a ΔE change that was smaller with respect to the other three samples. These two pigments have a similar chemical composition, as they both contain Zr. Zirconium oxide is used in a huge variety of applications, such as in optical layers and pigments [27], because of its superior chemical and optical properties. It has a high index of refraction, optimal, chemical, and thermal stability, in addition to the fact that it has been demonstrated as increasing the photostability of the TiO_2 layer [28,29]. The presence of ZrO_2 could increase the photostability of the pigments, thus limiting their degradation and aesthetical property changes. The variation of ΔE is mainly due to a variation of brightness, L^*, which follows a sigmoidal trend as well. Making a comparison with the colour variation and the gloss trend, shown in Figure 3, it is possible to observe that the degradation process started after 300 h of UVB irradiation. The a^* parameter was included between 0.1 and −0.35; the LG and GB-SW pigments showed quite a negative variation. Regarding the b^* parameter, it varied from 0.10 to −0.50; the only sample which showed a positive increment of b^* was the GB-SW sample, whereas the LG pigment showed a major change of the b^* coordinate.

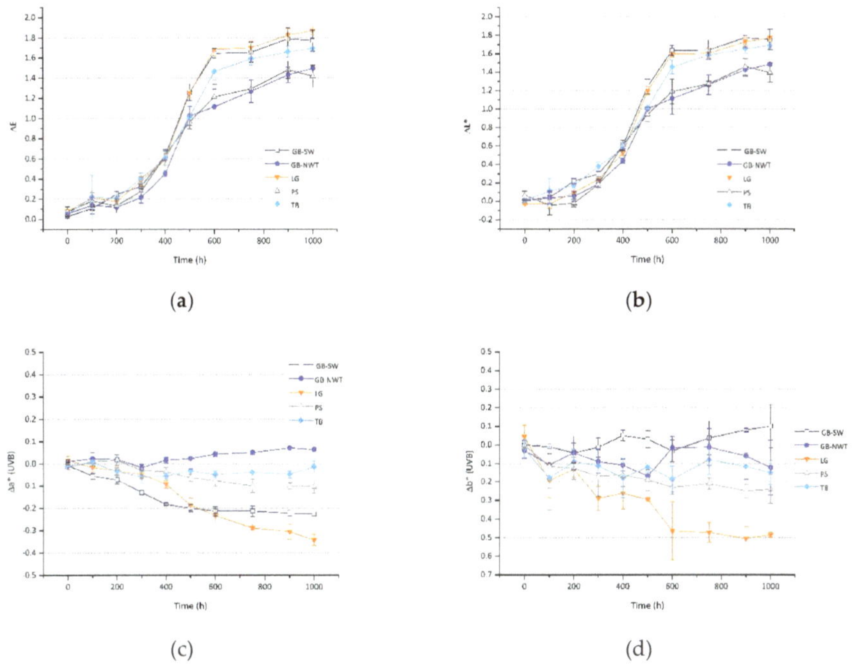

Figure 5. Colorimetric coordinates variation after UVB light exposure: (**a**) ΔE; (**b**) ΔL^*; (**c**) Δa^*; and (**d**) Δb^*.

3.4. FT-IR Analysis after UVA and UVB Exposures

In order to investigate the degradation and durability of powder coatings, FT-IR analyses were performed. Powder coatings based on polyesters, and in particular based on IPA, are used to withstand weathering, and they show an increased weathering resistance with respect to other coatings [30,31]. Different degradation mechanisms can affect the durability of superdurable resins [32], though the photo-induced oxidation mechanism is the most powerful and is mainly caused by resin absorbing UV radiation. The wavelengths between 290 and 370 nm cause important damages to the coatings because they lead to a change in the chemical structure of the polymeric matrix. Figure 6 shows the FT-IR spectra of the GB-SW sample before and after 1000 h of UVA exposure. It is possible, as a first thing, to assign the main peaks of the spectra.

A strong characteristic band associated with the carbonyl C=O stretching could be identified at 1715 cm^{-1} [31]. The peak at 1652 cm^{-1} can be associated to the C=O stretching of the curing agent typically used in this type of coatings [32]. The peak at 1608 cm^{-1} is related to the –CH– isophthalic bond stretching of the aromatic ring present in the isophthalic acid. The peaks at 1300 cm^{-1} and 1220 cm^{-1} are typical of C–O stretching. The peak at 1473 cm^{-1} could be both the bending of the $-CH_2-$ aliphatic group and the C=O bending; the peak at 1373 cm^{-1} corresponds to the CH_3 bending. The peak at 1072 cm^{-1} is related to the bending of aromatic C–H. No differences between the sample before and after UVA exposure could be detected. The same behaviour was followed by all the other samples, thus indicating that UVA radiation did not have an appreciable effect on the powder coating degradation.

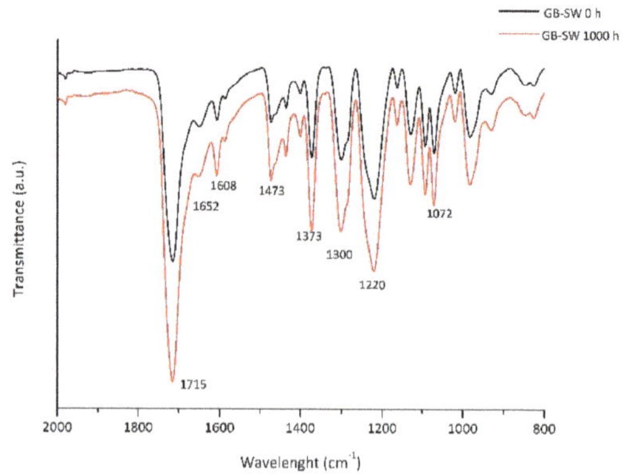

Figure 6. FT-IR spectra of the GB-SW sample before and after 1000 h of UVA exposure.

Instead, regarding the results obtained for the samples after UVB exposure, some differences could be noticed. Figure 7 shows the FT-IR spectra before and after 1000 h of UVB exposure.

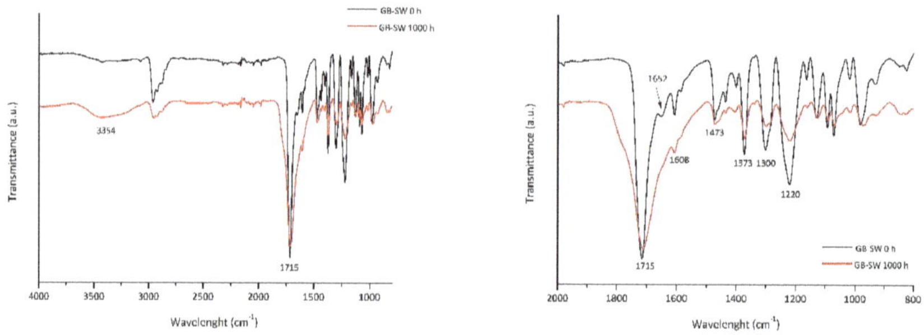

(a) (b)

Figure 7. FT-IR spectra of the GB-SW sample before and after 1000 h of UVB exposure: (**a**) reference spectra; (**b**) close-up of the (**a**) image.

Most of the peaks are shifted to lower wave numbers and have a lower transmittance, mainly due to the increase in mobility of the bonds with the absorption of UVB radiation. The two typical peaks of polyester at 1715 and 1652 cm^{-1} were different after 1000 h of exposure. The first peak underwent a shift towards 1711 cm^{-1} and a considerable broadening, whereas the latter almost disappeared. After UVB irradiation, a large band at 3354 cm^{-1} appeared, which could be associated with the –OH stretching vibration of hydroxyl groups, which may originate in the terminal phase of the degradation mechanism, already explained by Maetens [32]. The same behaviour was followed by all the five-sample series. It is then interesting to evaluate the FT-IR spectra of the samples at short exposure times, as shown in Figure 8.

The peak at 3354 cm^{-1} appeared after 300 h irradiation, while at the same time the peak at 1652 cm^{-1} disappeared. This change is the evidence of the initiation of the degradation process. All the samples followed the same trend shown for the GB-SW sample. A photo-induced oxidation mechanism generated by a photo-inductor can explain these results [32]. The samples exposed to UVB light underwent a degradation process starting from 300 h of irradiation, which can be correlated to the sample colour change shown in Figure 5.

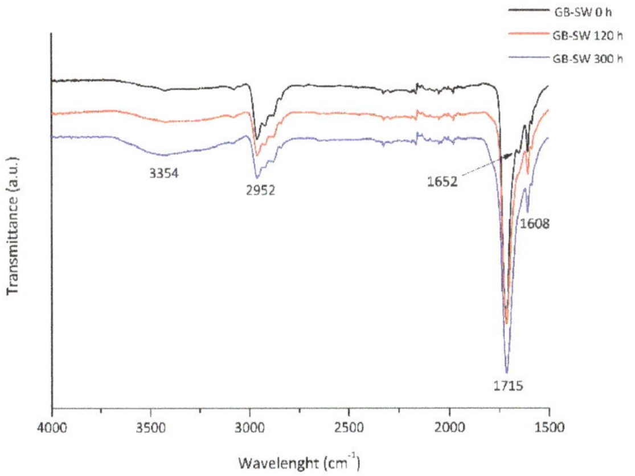

Figure 8. FT-IR spectra of the GB-SW sample before exposure, after 120 h, and after 300 h of UVB irradiation.

3.5. Exposure to an Aggressive Environment: Acetic Salt Spray Exposure

In order to determine the protective properties of the coatings and assess the adhesion of the coating with respect to the substrate, the samples were exposed to acetic salt spray fog for 1000 h in the presence of an artificial defect. As it is possible to observe in Figure 9, all the samples showed the presence of blisters of grade 4 (S5) [33], limited to the borders and the area close to the artificial defect. In addition to that, all the samples showed no blistering until 650 h of exposure, underlining the good protection properties of the coatings. In any case, it is important to point out that the evaluation of the blistering grade should be done on as-made samples, without the presence of an artificial scratch: considering the area far from the defect, no blisters could be observed.

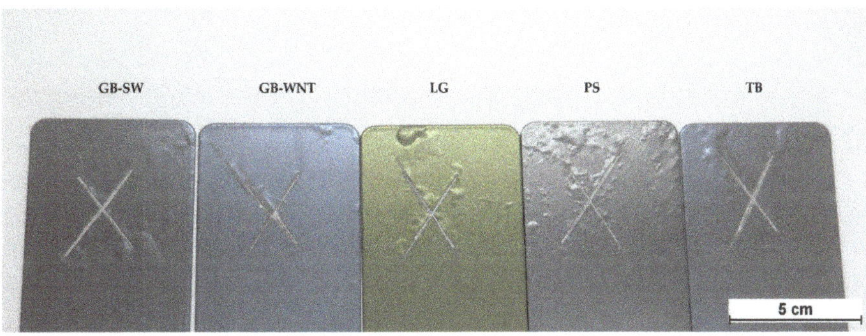

Figure 9. Samples after 1000 h exposure in an acetic salt spray chamber in the presence of an artificial scratch.

A delamination and corrosion assessment was carried out on the sample in presence of the artificial scratch, according to the UNI EN ISO 4628-8 standard [20]: the values obtained for delamination were included between 4.0 and 8.0 mm, the values obtained for corrosion were about 5 mm for all the samples.

Only small differences among the different samples could be observed. The slightly different thicknesses of the coatings does not represent an influencing factor since in this case the mechanism that could lead to a loss of adhesion between the coating and the substrate was represented by anodic undermining (typical of painted aluminium alloys), which is not influenced by the coating thickness, unlike what occurs in cathodic delamination [34]. The substrate pre-treatment could influence the adhesion of the coating on the substrate [35], then leading to the formation of blisters, but it is difficult to add this motivation to explain little differences between samples. On the contrary, these differences could be explained by little fluctuations of temperature or humidity in the deposition processes, which can have a negative influence on the adhesion with the substrate. It is necessary to underline that no relevant studies regarding the effect of aluminium-based pearlescent pigments on the degradation of powder coatings can be found in the literature, but it is reasonable to state that the different pigments have no influence on the degradation of the samples.

In order to have further insights on the protection properties of the coatings, EIS measurements were carried out before and after 400 h, 750 h, and 1000 h of exposure in an acetic salt spray chamber on as-made samples. In Figure 10, the Bode diagram of the GB-SW sample is shown. It is possible to observe that the impedance modulus |Z| at 10^{-2} Hz remained stable in time with a value of around 10^{11} $\Omega\cdot cm^2$, typical of protective organic coatings. Thus, the coating can be considered highly protective also after exposure to an aggressive environment.

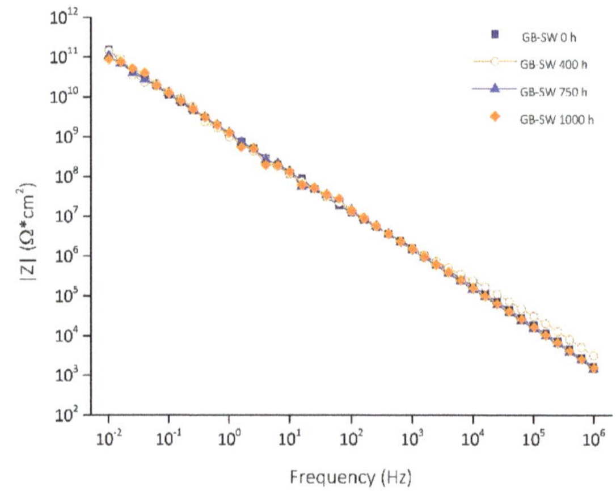

Figure 10. Bode diagram of the GB-SW sample at different acetic salt spray exposure times.

All the samples show a similar behaviour, and the impedance modulus |Z| versus applied frequency follows the same trend. Table 4 reports the values of |Z| at 10^{-2} Hz after 0, 400, 750, and 1000 h of acetic salt spray exposure for all the samples under investigation.

Table 4. Impedance modulus |Z| (Ω*cm^2) at 10^{-2} Hz after 0, 400 h, 750 h, and 1000 h.

Coating Name	0 h	400 h	750 h	1000 h
GB-SW	1.56×10^{11}	1.43×10^{11}	1.06×10^{11}	8.66×10^{10}
GB-WNT	1.57×10^{11}	1.06×10^{11}	8.66×10^{10}	8.40×10^{10}
LG	1.66×10^{11}	1.38×10^{11}	9.25×10^{10}	8.67×10^{10}
PS	1.23×10^{11}	8.67×10^{10}	6.70×10^{10}	6.23×10^{10}
TB	1.59×10^{11}	1.38×10^{11}	9.58×10^{10}	8.69×10^{10}

All the samples, as shown in Table 4, maintain a very high value of the impedance modulus, also after 1000 h of acetic salt spray exposure. Thus, it is possible to confirm that all the samples, without the presence of surface defects, can guarantee optimal protection properties, regardless of the used pigment.

4. Conclusions

This work examined the effect of five different pearlescent pigment types on standard powder coatings. The coatings' aesthetical and protective property changes were analysed as a consequence of accelerated ageing testing. The colour, gloss, and roughness were evaluated after UVA and UVB exposure. Regarding the samples subjected to UVA radiation, the surface colour did not register important changes, or, in any case, they were imperceptible to the human eye. Even the variation in gloss and roughness that was recorded was not significant. The results obtained after UVB exposure showed a different trend: the surface roughness showed a general increase, closely related to the decrease of gloss and the variation of colour. The decreasing gloss variation can be considered acceptable from a commercial point of view, while the colour change is very huge and could be attributable to the degradation of the polymeric matrix. The FT-IR analysis showed that UVB radiation caused a photoinduced degradation of the polymeric matrix, whereas the UVA radiation had no influence on the degradation of the coating. The degradation mechanism, due to UVB radiation, started after 300 h of exposure, and this is closely related to the abrupt change of colour and gloss. The analyses carried out after exposure to an acetic spray fog showed the good protection properties of the coating. EIS measurements showed that all the samples remained highly protective, also after 1000 h of acetic salt spray chamber exposure. In addition to that, all the samples showed no remarkable presence of blisters and detachment of the coating. On the other hand, a different behaviour in the presence of an artificial defect could be observed: a quite severe delamination and corrosion could be noticed. In conclusion, it is possible to state that these samples are very sensitive to UVB radiation but show good resistance to UVA light and to aggressive environments without the presence of coating defects. No pigment effect on the degradation of the samples could be revealed, though some differences in colour changes could be seen among the different sample types, which can be attributed to changes in the chemical composition.

Author Contributions: Project Administration, S.R.; Supervision, S.R.; Investigation, L.B.R.; Formal Analysis, L.B.R., F.R.; Data Curation, F.R.; Writing–Original Draft Preparation, F.R.; Writing–Review, S.R. All authors have read and agreed to the published version of the manuscript.

Funding: This research received no external funding.

Acknowledgments: The authors gratefully acknowledge Akzo Nobel Coatings S.p.A. (Como, Italy) in the person of Roberto Paganica and Merck Italia S.p.A. (Milano, Italy) in the person of Stefano Corrado for supplying the coated samples and the pigments, respectively.

Conflicts of Interest: The authors declare no conflict of interest.

References

1. Ashby, M.F.; Johnson, K. *Materials and Design: The Art and Science of Material Selection in Product Design*, 2nd ed.; Butterworth-Heinemann: Oxford, UK, 2009.
2. Rossi, S.; Deflorian, F.; Scrinzi, E. Reduction of aesthetical properties of organic coatings caused by mechanical damage. *Mater. Des.* **2009**, *30*, 1511–1517. [CrossRef]
3. Perera, D.Y. Effect of pigmentation on organic coating characteristics. *Prog. Org. Coat.* **2004**, *50*, 247–262. [CrossRef]
4. Pfaff, G.; Franz, K.; Emmert, K.; Nitta, R. Pigments, Inorganic. In *Ullmann's Encyclopedia of Industrial Chemistry*, 6th ed.; Wiley-VCH: Weinheim, Germany, 1998.
5. Bauxbaum, G.; Pfaff, G. *Industrial Inorganic Pigments*, 3rd ed.; Wiley-VCH: Weinheim, Germany, 2005. [CrossRef]
6. Jansen, M.; Letschert, H.P. Inorganic yellow-red pigments without toxic metals. *Nature* **2000**, *404*, 980–982. [CrossRef] [PubMed]
7. Wendusu, T.; Masui, T.; Imanaka, N. Novel environmental-friendly inorganic red pigments based on (Bi, Er, Y, Fe)$_2$O$_3$ solid solutions. *J. Asian Ceram. Soc.* **2014**, *2*, 195–198. [CrossRef]
8. Faulkner, E.B.; Schwartz, R.J. *High Performance Pigments*, 2nd ed.; Wiley-VCH: Weinheim, Germany, 2009. [CrossRef]

9. Maisch, R.; Weigand, M. *Pearl Luster Pigments, Physical Principles, Properties, Applications*; Verlag Moderne Industrie: Landsberg, Germany, 1992.
10. Pfaff, G. *Special Effect Pigments: Technical Basics and Applications*; Vincentz Network GmbH & Co KG: Hannover, Germany, 2008.
11. Maile, F.J.; Pfaff, G.; Reynders, P. Effect pigments – past, present and future. *Prog. Org. Coat.* **2005**, *54*, 150–163. [CrossRef]
12. Maisch, R.; Stahlecker, O.; Kieser, M. Mica pigments in solvent free coatings systems. *Prog. Org. Coat.* **1996**, *27*, 145–152. [CrossRef]
13. Pfaff, G.; Becker, M. Special effect pigments in cosmetic applications. *Househ. Pers. Care Today* **2012**, *1*, 12–15.
14. Pfaff, G. Special effect pigments based on silica flakes. *Inorg. Mater.* **2003**, *39*, 123–126. [CrossRef]
15. *ASTM G154-16 Standard Standard Practice for Operating Fluorescent Ultraviolet (UV) Lamp Apparatus for Exposure Non-Metallic Materials*; ASTM International: West Conshohocken, PA, USA, 2016.
16. *ASTM D523-14(2018) Standard Standard Test Method for Specular Gloss*; ASTM International: West Conshohocken, PA, USA, 2018.
17. *ISO 4288:1996 Standard Geometrical Product Specification (GPS)—Surface Texture: Profile Method—Rules and Procedures for the Assessment of Surface Texture*; International Organization for Standardization: Geneva, Switzerland, 2018.
18. *ASTM B117-19 Standard Standard Practice for Operating Salt Spray (Fog) Apparatus*; ASTM International: West Conshohocken, PA, USA, 2019.
19. *ASTM G85-19 Standard Standard Practice for Modified Salt Spray (Fog) Testing*; ASTM International: West Conshohocken, PA, USA, 2019.
20. *UNI EN ISO 4628-8(2013) Standard Evaluation of Degradation of Coatings—Designation of Quantity and Size of Defects, and Intensity of Uniform Changes in Appearance—Part 8*; UNI Ente Nazionale Italiano di Unificazione: Milano, Italy, 2013.
21. Topuz, B.B.; Gündüz, G.; Mavis, B.; Çolak, Ü. The effect of tin dioxide (SnO_2) on the anatase-rutile phase transformation of titania (TiO_2) in mica-titania pigments and their use in paint. *Dye. Pigm.* **2011**, *90*, 123–128. [CrossRef]
22. Gao, Q.; Wu, X.; Fan, Y. The effect of iron ions on the anatase-rutile phase transformation of titania (TiO_2) in mica and titania pigments. *Dye. Pigm.* **2012**, *95*, 96–101. [CrossRef]
23. Hosseini Zori, M. Particle size and kind of mica in synthesis of nontoxic bronze and gold pearlescent pigments based on nanoencapsulated hematite. *J. Ultrafine Grained Nanostruct.* **2015**, *48*, 101–112. [CrossRef]
24. *GSB AL 631 Standard, "International Quality Regulations for the Coating of Building Components—Aluminium"*; GSB International e.V.: Düsseldorf, Germany, 2017.
25. Hunter, R.S.; Harold, R.W. *The Measurement of Appearance*, 2nd ed.; John Wiley & Sons: Hoboken, NJ, USA, 1987.
26. *ASTM E308-18 Standard, "Standard Practice for Computing the Colors of Objectives by Using the CIE System"*; ASTM International: West Conshohocken, PA, USA, 2018.
27. Zhang, Q.; Shen, J.; Wang, J.; Wu, G.; Chen, L. Sol-gel derived ZrO_2–SiO_2 highly reflective coatings. *Int. J. Inorg. Mater.* **2000**, *2*, 319–323. [CrossRef]
28. Wei, B.-X.; Zhao, L.; Wang, T.-J.; Gao, H.; Wu, H.-X.; Jin, Y. Photo-stability of TiO2 particles coated with several transition metal oxides and its measurements by rhodamine-B degradation. *Adv. Powder Technol.* **2013**, *24*, 708–713. [CrossRef]
29. Mathivanan, L.; Arof, A.K. The effect of zirconium oxide and quartz pigments on the heat and corrosion resistance properties of the silicone based coatings. *Pigm. Resin Technol.* **2000**, *29*, 10–15. [CrossRef]
30. Rossi, S.; Fedel, M.; Petrolli, S.; Deflorian, F. Behaviour of different removers on permanent anti-graffiti organic coatings. *J. Build. Eng.* **2016**, *5*, 104–113. [CrossRef]
31. Gheno, G.; Ganzerla, R.; Bortoluzzi, M.; Paganica, R. Accelerated weathering degradation behaviour of polyester thermosetting powder coatings. *Prog. Org. Coat.* **2016**, *101*, 90–99. [CrossRef]
32. Maetens, D. Weathering degradation mechanism in polyester powder coatings. *Prog. Org. Coat.* **2007**, *58*, 172–179. [CrossRef]
33. *UNI EN ISO 4628-2(2016) Standard, "Evaluation of Degradation of Coatings—Designation of Quantity and Size of Defects, and Intensity of Uniform Changes in Appearance—Part 2"*; UNI Ente Nazionale Italiano di Unificazione: Milano, Italy, 2016.

34. Mirabedini, S.; Scantlebury, J.D.; Thompson, G.E.; Moradian, S. Adhesive strength of powder coated aluminium substrates. *Int. J. Adhes. Adhes.* **2005**, *25*, 484–494. [CrossRef]
35. Fedrizzi, L.; Stenico, M.; Deflorian, F.; Maschio, S.; Bonora, P.L. Effect of powder painting procedures on the filiform corrosion of aluminium profiles. *Prog. Org. Coat.* **2007**, *59*, 230–238. [CrossRef]

MDPI
St. Alban-Anlage 66
4052 Basel
Switzerland
Tel. +41 61 683 77 34
Fax +41 61 302 89 18
www.mdpi.com

Coatings Editorial Office
E-mail: coatings@mdpi.com
www.mdpi.com/journal/coatings